"Brimming with information and amply footn[...]k, [Bread, Wine, Chocolate is a] revelation throughout."
—*Science*

"... Full of wonderfully geeky bits of science, including an excellent section on how memory and culture influence our perception of taste. Sethi's friendly, welcoming tone makes serious topics digestible and pleasurable. 'Eat and drink with reverence and gusto, whether it's a Big Mac or a mountain of kale,' she writes, with an admirable lack of foodie pretension."
—Associated Press

"When we taste, we taste history—the interplay of crops, livestock, and even wild things—with our own human cultures. And we taste diversity: genetic diversity, cultural diversity. Simran Sethi's book opens this world to a new generation by focusing on foods we think we know, but don't. *Bread, Wine, Chocolate* helps us understand the richness of these foods and others, and why it is essential to preserve diversity if we wish to appreciate and fully benefit from such foods in the future. Readers of this book will both enjoy and be enlightened; many will even find their taste buds subtly changed by a new awareness of what they are really eating."
—Cary Fowler, senior advisor to the Global Crop Diversity Trust and author of *Shattering: Food, Politics, and the Loss of Genetic Diversity*

"Is biodiversity the key to a better cup of coffee? And how sexy can achieving food security really be? Simran Sethi's answers are 'yes' and 'you'd be surprised.' In her book, [she] looks at ways in which monoculture and an increasingly standardized global diet put food systems in peril and leave crops vulnerable to blight and climate change. And she does so winningly, by relishing her favorite things to eat and drink, visiting the places they're produced, digging up their stories, and teasing out nuances of flavor unique to individual varieties and landscapes."
—*The Wall Street Journal*

"Simran Sethi's passionate book on food and biodiversity reminds us how healing food can be. The world is on our plate."
—Deepak Chopra, M.D., author of more than 80 books, including *Super Genes*

"We need more investigative journalists like Simran Sethi. She writes with a deep understanding of pleasure and taste to convey her urgent message: We must make uncompromising, purposeful choices when it comes to what we eat before it's too late! *Bread, Wine, Chocolate: The Slow Loss of Foods We Love* is a loving call to action that we must heed."
—Alice Waters, chef, author, and proprietor of Chez Panisse

"This absorbing book serves to remind us that biodiversity—through a diversity of species, varieties, and breeds of plant and animal foods, including wild food—underpins dietary diversity, good nutrition, and health. It also underscores the need to place a renewed emphasis on sustaining the natural varieties of crops and animals contributing to agriculture, including neglected yet nutritious traditional foods, in order to improve food security."

—Braulio Ferreira de Souza Dias, Executive Secretary of the Secretariat, Convention on Biological Diversity

"By turns explorer and explainer, Simran Sethi conducts a thoughtful and heartfelt tour of humanity's most beloved tastes and the threats that could extinguish them forever."

—Michael Brune, Executive Director, Sierra Club

"In this illuminating and impactful book, Simran Sethi sheds light on the dwindling diversity of our diets and our landscapes through the stories of our most beloved tastes. The solution to this global agricultural and culinary crisis, she argues, lies in our collective palates. *Bread, Wine, Chocolate* calls on all of us to cherish—and thus preserve—the world's endangered flavors."

—Dan Barber, chef/co-owner of Blue Hill Restaurant and *New York Times* bestselling author of *The Third Plate*

"Unlike many other recent books that cover some of the same territory, what sets Sethi's work apart is her joyous, generous attitude toward the human appetite. Yes, she argues for the importance of the local, organic, and artisanal, but there's nothing pretentious, lofty, or hectoring in her tone. 'Eat and drink with reverence and gusto, whether it's a Big Mac or a mountain of kale,' Sethi writes, is an invitation to 'the slow savor' that bonds eaters to food and earth."

—*The Boston Globe*

"An important component of agrobiodiversity is culinary and cultural diversity. This beautiful book brings into sharp focus the threat to culinary diversity as a result of genetic erosion. Already the food basket is shrinking, and a few crops are occupying the dominant position in our daily diet. Simran Sethi has, through this book, brought out clearly the importance of diversity of food and the need for the conservation of the foods whose tastes we cherish and enjoy. The book provides a rationale for the need for a culinary revolution in all parts of the world."

—M. S. Swaminathan, Founder, M. S. Swaminathan Research Foundation

"A powerful reminder that we can eat in ways that don't cause damage to the planet or its poorest people—and that can delight us, not just fill us up. Don't read it on an empty stomach!"

—Bill McKibben, Schumann Distinguished Scholar in Environmental Studies at Middlebury College and *New York Times* bestselling author of *Eaarth: Making a Life on a Tough New Planet*

"In *Bread, Wine, Chocolate: The Slow Loss of Foods We Love*, Simran Sethi sets out to discover diversity—to find the corners of the world where the foods we cherish are thriving in multiplicity instead of surviving in monoculture. Sethi travels across six continents to talk to farmers, brewers, bakers, and winemakers working to foment more delicious food and drink and, in doing so, build resilience in agriculture."

—*Orion*

"Read this wonderful book and you will become immersed in the intricate worlds of no less than six (delicious) foods and drinks. But this is not really a book about food. Rather, it is about our relationships with the life forms that sustain us—and how we might learn to approach those relationships with far more love, compassion, and good taste."

—Naomi Klein, *New York Times* bestselling author of
This Changes Everything and *The Shock Doctrine*

"A world in which we all consume the same food will end up with a serving of disaster. Supermarket shelves filled with the exotic give a false impression, warns environmentalist Simran Sethi. Her new book exposes the dangers of eating a small number of the same things."

—*The Guardian*

"Our tables ... are never really of—or for—one, as Sethi elegantly shows us."

—National Public Radio

"A heartfelt lament for the homogenization of our taste buds."

—*The Independent*

"Follow Simran on a journey with a lifetime companion: taste buds. They are not baubles to be manipulated by food conglomerates. Taste buds are the most precious of gifts. They are evolution itself, a teacher, a kindness, a guide sorting out millions of molecules, a doorman making sure that everything we chew and savor is on the guest list.

In this needed and nourishing volume, you will discover how tastes can recover an even larger sense, a self that is intimately connected to a vast, pulsating life of ecstasy and delight. Read this and you will understand that cuisine is how we kiss the world. There is more good news: It kisses back."

—Paul Hawken, *New York Times* bestselling author of
Natural Capitalism and *Blessed Unrest*

"A passionate plea to save and restore the things most precious about our food— its myriad flavors and its connection with nature. As global economic forces slowly squeeze the uniqueness out of what we eat, Simran Sethi explores the culinary delights that offer hope, and deliciousness, for the future."

—John McQuaid, Pulitzer Prize-winning journalist and author of
Tasty: The Art and Science of What We Eat

"This book is the *Eat, Pray, Love* of the diversity of food. The history of how bread, wine, and chocolate entered our lives, related by Simran Sethi, is both inspiring and humbling. Humbling because of the richness and diversity of tastes that we can enjoy and that fill our souls. And inspiring, as it invites us to be curious about their origins and mindful of the hard work of the many men and women that grow, process, and make these delights possible. The diversity is threatened and disappearing fast, and this book wakes us up to value these precious resources and stop this process."

—Brigitte Laliberté, Bioversity International

"While she maps the geography of food, Simran has also mapped the geography of the heart."

—*University of California Food Observer*

"*Bread, Wine, Chocolate: The Slow Loss of Foods We Love* is destined to join the books of Michael Pollan and Dan Barber as must reads for anyone interested in biodiversity and sustainability. Her book, loaded with important information about the fragility of our ecosystem and food chain, is also lush with great food writing and tasting strategies for how anyone can access the subtle flavors and regional distinctiveness of chocolate, coffee, beer, wine, and, of course, bread. More importantly, while the book is alarming as it reveals how close we are to wiping out vital species ... it also provides hope and a way that each of us can help effect a transformative change."

—Peter Reinhart, baker and James Beard Award-winning author of
The Bread Baker's Apprentice

"When Simran Sethi lasers in on a topic, stop and take notice. She is a fierce observer and a cool head in a world riven with misinformation, disinformation, and just plain wrong-headedness. Her new book, *Bread, Wine, Chocolate*, should be required reading for culinary students, chefs in training, journalists, scholars, cooks, and citizens who care about what they put into their mouths, as well as what we're doing to Mother Earth. All I can say to you is there's a new expert coming on the scene. A hot new voice of reason in a cold, frightened world, Simran Sethi is the kind of writer who can coat the bitter pill in honey, and we all just swallow and say thank you."

—Linda West Eckhardt, IACP Julia Child Award-winning coauthor of
Bread in Half the Time

BREAD
WINE
CHOCOLATE

The Slow Loss of Foods We Love

SIMRAN SETHI

HarperOne
An Imprint of HarperCollinsPublishers

For every person listed in the acknowledgments.
Thank you for feeding and nourishing me.
This *is* because of you.

For Mom and Dad.
Your DNA. Your best beti.
I am because of you.

For my nephews, Dev and Avi.
May you inherit a delicious and resilient world
where all have a seat at the table.

HarperOne

"Love After Love" from *The Poetry of Derek Walcott* by Derek Walcott, selected by Glyn Maxwell. Copyright © 2014 by Derek Walcott. Reprinted by permission of Farrar, Straus and Giroux, LLC.

BREAD, WINE, CHOCOLATE. Copyright © 2015 by Preeti S. Sethi. All rights reserved. Printed in the United States of America. No part of this book may be used or reproduced in any manner whatsoever without written permission except in the case of brief quotations embodied in critical articles and reviews. For information address HarperCollins Publishers, 195 Broadway, New York, NY 10007.

HarperCollins books may be purchased for educational, business, or sales promotional use. For information please e-mail the Special Markets Department at SPsales@harpercollins.com.

HarperCollins website: http://www.harpercollins.com

FIRST HARPERCOLLINS PAPERBACK EDITION PUBLISHED IN 2016

Designed by Kris Tobiassen of Matchbook Digital

Library of Congress Cataloging-in-Publication Data is available upon request.

ISBN 978-0-06-158108-3

16 17 18 19 20 RRD(H) 10 9 8 7 6 5 4 3 2 1

LOVE AFTER LOVE

The time will come
when, with elation,
you will greet yourself arriving
at your own door, in your own mirror,
and each will smile at the other's welcome,

and say sit here. Eat.
You will love again the stranger who was your self,
Give wine. Give bread. Give back your heart
to itself, to the stranger who has loved you

all your life, whom you ignored
for another, who knows you by heart.
Take down the love-letters from the bookshelf

the photographs, the desperate notes,
peel your own image from the mirror.
Sit. Feast on your life.

—DEREK WALCOTT

CONTENTS

INTRODUCTION

This is a book about food, but it's really a book about love. It's about that moment when you find yourself savoring something so wholly and intently you never want to let it go. I thought this love, at least in the culinary sense, could only be found in superlative places: a secret supper club in London, a hidden bistro in Paris or a roadside *dhaba* in Mumbai. But I know now the greatest love is found in humble places: in my morning coffee, in a morsel of bread or in a bite of chocolate. And that to pay closer attention to these ordinary pleasures isn't just to see them anew but to experience them in a whole new way.

I had forgotten how to do this. I had forgotten how to be present to what was right in front of me, knowing only how to love what shouted for my attention. Until I realized I could lose them.

I spent the spring of 2012 researching this book in Rome, Italy, where the saying "When in Rome" took on a life of its own. When in Rome, start the day with *un caffè e una sigaretta*! When in Rome, *mangia un gelato* every afternoon! When in Rome, start drinking at five: *In bocca al lupo*!

Four months later, I returned to the United States chubby and tired, primed for a cleanse. And primed I was as I walked through San Francisco's Embarcadero to the headquarters of chocolate maker TCHO (pronounced "cho"), basking in the virtuous glow of the no-sugar-no-dairy-no-gluten-no-alcohol-or-cigarettes-or-anything-that-could-be-construed-as-sinful cleanse I had started days before I was to interview the company's head chocolate maker, Brad Kintzer.

I stepped into the chocoholics' lair and asked for Brad. (If you ever eat a TCHO chocolate bar, you'll find a photo of him smiling beatifically on the wrapper's inner fold.) About five minutes later, he walked out, apologized for running late and requested another 15 minutes to finish his work. He invited me to order a cup of hot chocolate from the café to pass the time and sweeten the delay: "Order whatever you want." I thanked him, waited until he left and ordered a cup of water. I was cleansing—and virtuous. So virtuous.

To be inside the original TCHO space (they have since moved) was akin to placing myself inside a Willy Wonka dream: Simi and the Chocolate Factory. Brad brought me into the conference room and explained they were putting the finishing touches on a new hazelnut bar. The room was heady with the aromas of nuts and chocolate; broken samples were scattered all over the conference table. "Help yourself," he said. I smiled, beatifically. "I'm okay. Thanks."

The 20-minute interview stretched to almost two hours. I was captivated by Brad's story, his journey from a man who had started off studying the biology of sugar maples to now making award-winning chocolate. At one point, he described the moment he shared some of that chocolate with the farmers who'd grown the cacao—men who had never before tasted a finished chocolate bar—as "one of the most sacred moments in all my life."

I was starting to regret my cleanse.

As we wrapped up the interview, Brad asked if I wanted to tour the factory. *Of course.* Willy Wonka was giving me a tour of his chocolate factory. Brad and I slipped on mesh hairnets, smooshed in orange earplugs and walked into the outer perimeter of the factory where bars are molded and hand wrapped. It was chilly; the area is kept below 66 degrees Fahrenheit to maintain the consistency of the chocolate. And it was loud; Brad shouted over slappers that hit chocolate bars out of their molds, plus cooling and packing machines that churned out roughly 5,000 TCHO bars per hour.

He then pushed through a thick plastic curtain and led me into the inner sanctum, a cozier 80 degrees Fahrenheit. The aroma of chocolate grew stronger as we approached the refiner, a machine that grinds and melts solid cakes of cocoa into a warm, gooey mass.

That's where I nearly buckled.

As Brad explained the transformation of solid into liquid, I closed my eyes. The scent of chocolate was so overwhelming, my mouth started to water. "The fat in chocolate is solid at room temperature," he said. I swallowed; I could taste the chocolate without tasting it. "Chocolate melts just below the temperature of our mouths."

I caressed the refiner as if I were touching a lover. The drum was so warm, the smell so intoxicating. Brad was shocked. He stopped mid-sentence and asked if he could take a picture. I was still in my hairnet, covered in various shades of brown—brown skin, brown jacket, brown bag, brown boots—looking nearly post-orgasmic. I was embarrassed, but Brad understood. He smiled and said, "We're born loving chocolate."

He's right. Taste preferences are established in utero. Our first taste buds develop eight weeks after conception. At 12 weeks, when a fetus begins swallowing, the smells in amniotic fluid ignite its taste receptors. When we're born, taste is our primary and most developed sense—an evolutionary response to help us steer clear of poison. Sweetness (the dominant quality in mothers' milk) signals the presence of carbohydrates, safe sources of energy. Bitterness warns us of toxicity, which babies—and prehistoric humans—are programmed to avoid.[1]

Taste is not only a function of biology. It's also shaped by what our mothers ate during pregnancy. Infants show a predisposition toward certain foods consumed in utero, months before birth. Throughout my mother's pregnancy, she ate ginger (for morning sickness) and chocolate (for pleasure). I was born with an appreciation for both.

Chocolate has been my constant companion: every birthday cake, my wedding cake, the food that got me through my divorce. It, along with coffee and the occasional cigarette, has fueled every single page of this book. But despite this love, I had never thought deeply about where it came from—or where any of my favorite foods came from—beyond a fuzzy notion of "farmers in fields" and "workers in factories." They were people whom I considered in the abstract but did not know.

Yes, I'm friendly with the farmers who sell me eggs and seasonal produce at my local farmers' market, but most of the people who cultivate what *I* consider life staples (including chocolate and coffee) don't even live on my continent. Despite my passion for food and agriculture, and my deep care for land and people, my relationships with the foods I love have been long but not deep.

I didn't spend most mornings thinking of where my coffee came from.

I didn't spend *any* mornings thinking about where my coffee came from.

Now I do.

Because coffee, chocolate, bread—every food we care about—is under threat. While we debate GMOs and the merits of Paleo, while we count calories and queue for Cronuts, we're losing the foundations of food. That's what I learned when I traveled to Rome to research challenges in modern agriculture. Embedded in every conversation about feeding people, conserving natural resources and ensuring a healthy diet, both now and in the future, is the threat of the loss of agricultural biodiversity—the reduction of the diversity in everything that makes food and agriculture possible, a shift that is the direct result of our relationship with the world around us.

Once I learned about this, I knew I had to go to the places that hold the keys to the future of food. So I quit a job I couldn't get fired from, sold my house, gave away my car and embarked on a journey to learn how we could save the tastes we love.

This is a book about love, but it's really a book about taste. I have spent over three years meeting tireless, courageous and innovative people dedicated to making our food supply secure, abundant and more delicious, traveling across six continents to interview more than 200 scientists, farmers, chefs, bakers, winemakers, beer brewers, coffee roasters, chocolate connoisseurs, conservationists, religious leaders, and advocates and experts of all types to learn the intimate stories of our foods and ways we can better save—and savor—them.

In order to do this, we have to go deep—to the origins. Every food has an inspiring birthplace and holds flavors directly connected to the places and people that make them. And every food is under threat. Once we learn to recognize and appreciate these differences, our experience of what we eat will change and, subsequently, so will the system that creates our food.

The solutions are in the places I will take you: the Ethiopian coffee forest, the British yeast cultures lab, the vineyards of California, the cacao plantations of Ecuador, the brewery, the bakery and the temple. And they are in you—in us. "Taste" is a noun and a verb: We all have it and we all do it. But we don't all have a language or system for understanding and expressing that experience. In that moment of caressing the refiner at TCHO, I knew chocolate was something I didn't want to lose, but I didn't have the words to communicate why it was so important to me, or the knowledge on how best to save it.

Now I do.

And through this journey—through wine, chocolate, coffee, beer and bread—you will, too. You'll deepen your pleasure and understanding of what you eat and drink even if you're a gluten-free, sugar-free, caffeine-free, vegan teetotaler. Because everything I learned about my culinary staples can be mapped onto yours.

The changes we will explore are a reflection of what's happening to all foods and drinks, but only one of the crops I included—wheat—is what scientists and nutritionists would call a dietary staple. Staples like rice, corn and wheat make up over two-thirds of the world's diet. But they aren't what get me out of bed in the morning or help me celebrate at night. If you've ever raised a glass for a toast or woken up to Folgers in your cup, you know how meaningful so-called non-staples can be. They're the stuff of life and love—celebrated, debated and imbued with far more than calories. For the Chinese, tea drives the day, not coffee. For Peruvians, potatoes are both a nutritional and a soul staple. Rwandans grow bananas for beer as well as breakfast. Hindus revere cows for milk, not meat. Our menus are different, but our needs—ones that include and transcend nutrition—are universal.

The world, or at least every expert on dietary cleanses, tells us to be more ascetic, to eat more kale and drink more kombucha. That's great.

But I do not think it should be done at the expense of soul nourishment. We aren't on this planet to merely survive; we're here to take it all in and *thrive*. So rather than deprive ourselves of this joy, we should maximize it. If we're lucky enough to have any choice in what we're able to eat—to be able to eat at all—then we should honor this privilege by eating less of the bad stuff and more of the good, celebrating the fact that the solutions to the loss of agricultural biodiversity aren't difficult; they're delicious.

Until I embarked on this journey, I didn't understand that greater nourishment and a deeper savoring of every aspect of my life were not only available but what I deserved. What we all deserve—and can have. Not just crumbs of life, but cake.

"Eating," author, farmer and philosopher Wendell Berry says, "is an agricultural act."[2] Food connects us to all living things, and to the lineage of who we are and where we come from. It isn't farmers in fields and workers in factories who bring us our food; it's people like us. People who dedicate their lives to creating something that we take into our bodies. They transform nature into culture, as what they touch becomes part of us. This intimacy is astonishing and humbling.

We treat our food system as an abstract thing; however, it's a dynamic entity made up of these relationships, ones I have come to fully appreciate through the journeys on these pages. But this isn't where I started. The first wine I loved was a peach Bartles & Jaymes wine cooler. The first coffee I tolerated—heavily diluted with half-and-half and sugar, and chased with a glazed doughnut—was at the world's original Krispy Kreme. My favorite chocolate bars were Whatchamacallit, Twix and Nestlé Crunch—in that order. Until a year and a half ago, all I knew of coffee was that I preferred a cappuccino to a latte. But now I grill baristas about coffee origins, beg friends to bring back chocolate from various countries and quiz breweries about the source of their hops.

I am not trying to be precious. I have learned, by traveling to the places where some of our favorite foods and drinks began, that these foods *are* precious. I had no idea how hard it was to get a coffee bean from a forest in Ethiopia to my local café, or how much work and care went into making a premium bar of chocolate or a hearty loaf of bread.

I had no idea how endangered the best, most delicious versions of these things are. This awareness is what makes them precious, and, meal by meal, makes my life better.

By helping you more deeply understand foods that are already a part of your life, and develop a sensory map for exploring new ones, this book will enable you to discover and appreciate flavors you may not have experienced. It will give you the tools to define your own deliciousness—and reach for more.

That's been this journey's greatest reward: finding a new appreciation for what I already loved. And understanding what it takes to sustain and save that love—in farms, on our plates and in life—lies in the recognition that how we eat is a reflection of how we live. By sustaining agricultural biodiversity, we sustain ourselves.

"Eating with the fullest pleasure—pleasure, that is, that does not depend on ignorance," Berry adds, "is perhaps the profoundest enactment of our connection with the world. In this pleasure we experience and celebrate our dependence and our gratitude, for we are living from mystery, from creatures we did not make and powers we cannot comprehend."[3]

This is why people like Brad Kintzer devote their lives to making foods like chocolate taste better. "The more I looked into cacao and chocolate," he told me, "the more I realized it was an amazing way to see the world and understand how it really works. You can sing the virtues of bourbon all day, but it's not something that takes you all the way back. Chocolate always makes us happy."

This is a book about taste, but it's really a book about joy.

WHAT'S AT STAKE

I. BIODIVERSITY

Shortly after my tour of the TCHO factory, I decided to reward myself for being virtuous and *not* breaking my cleanse with chocolate ... by breaking my cleanse with wine. To me, this decision made perfect sense. Wine wasn't my vice. Although I had consumed an ungodly amount of it in Italy—some insanely delicious—it wasn't something I sought out or missed when it was gone. It wasn't something that had taken hold of me.

Until it did.

The waitperson at Camino restaurant in Oakland, California (whom, I later learned, is named Ali Hamerstadt), asked me if I wanted to try something different. "Sure," I said. "Surprise me." She was warm and pretty and seemed intent on bringing me something lovely, so I stepped out of my Chardonnay comfort zone and went for it. Ali returned with a glass of Trousseau Gris, a wine the color of pale gold. She told me the wine director for the restaurant had been introduced to the wine by the winemaker, and that the vineyard and the maker weren't too far from the restaurant.

I was intrigued. The wine was no longer just a way to unwind; it was a story culminating in the glass. I wanted to know more. But curiosity wasn't my norm. I hate unpredictability. I'm the type who looks at a menu online and decides what to order before I get to a restaurant. I am also loyal to a fault: When I find what I love, I stay with it. I mean this in every sense of the word.

In regard to food, I'm not alone. The standard American diet is, with a few notable exceptions, a supersized version of what we ate 40-odd years ago, made up of mostly grains, fats, oils and animal-based proteins.[1] We eat about the same amount of fruit today that we did in the 1970s (60 pounds a year) and the same amount of vegetables we ate in the 1990s (110 pounds). In the last 45 years, our milk consumption has dropped from 21 gallons to 13 gallons, but we moved the fat we used to get from whole milk over to cheese, which is why our dairy consumption has nearly tripled—from 8 pounds to 23 pounds per person. Our love of cheese has contributed to a whopping 20 additional pounds in total fat we eat each year.

We find what we love and stick with it—but that doesn't necessarily mean it's good for us. In its best manifestation, food *is* love—one of the most intimate connections that exists between people. But love is hard, and improving our relationships is work. It requires not only a commitment to ourselves and the objects of our care but a willingness to see and do things differently. In order to transform our love lives, change our diets or increase biodiversity, we first have to understand the connections and factors that inform those choices.

Agricultural biodiversity—also called "agrobiodiversity"—is the foundation of agriculture and food. It's what emerges out of the connection between:

1. the microorganisms, plants and animals we eat and drink;
2. the inputs that support their creation and development, including bees and other pollinators, as well as the quality of nutrients in the soil;
3. nonliving (or abiotic) influences on our ability to grow and gather food, such as temperature and the structures of farms; and
4. a range of socioeconomic and cultural issues that inform what and how we eat.[2]

Broadly speaking, biodiversity is the variety of life on Earth. (You'll find a reference guide in Appendix I.) "It comes in three levels: ecosystem, species and genetic," explains Luigi Guarino, senior scientist at the Global Crop Diversity Trust. (The Trust provides funding for important crop collections and oversees the Svalbard Global Seed Vault—what's popularly known as the "Doomsday Seed Vault" or "Noah's Ark" for seeds.) The first time I met Luigi, I was researching the loss of diversity in seeds. He was smart, funny and tough, reminding me, as he picked his teeth with my business card, that seeds are essential. "Seeds," he said almost dismissively, "are sex." *Indeed.* I blushed, not because it was salacious but because it was obvious. Luigi has dedicated his professional life to biodiversity and has, over the last few years, helped me understand (with a healthy dose of tough love) the stories of seeds and the intricacies of conservation.

The broadest level of biodiversity is that of ecosystems: the communities of plants, animals and all living creatures that interact with one another and their physical environment, such as a desert or a rainforest. The second level is species, defined as the largest group of individual organisms that can have sex and produce fertile offspring. Species exist *within* these ecosystems. The third level, genetic diversity, is measured in multiple ways and is tougher to discern. Genes are part of DNA (deoxyribonucleic acid)—the chains of molecules that are the instruction manuals for every cell in our body. "Counting differences in the sequences of our DNA is the best way to identify genetic diversity," Luigi explains. "But you can also count varieties, or measure variations in size, shape and structure ... what are known as morphological differences."

I was still unclear, so he elaborated: "It's not a great comparison, and probably politically incorrect, but think of how you would measure genetic diversity in 100 humans you picked off the street. DNA fingerprints, sure. But also, how many races and ethnicities are represented in the sample? That would be the very rough equivalent of varietal diversity. Finally, how many eye colors are there? And what is the range of heights? That's morphological diversity."

These influences are dynamic; they operate in response to one another and are constantly evolving. Agrobiodiversity shapes—and is shaped by—every meal we eat. And when I say "we," I mean "we" in the global sense, for all of us. We're all in this together: No country is self-sustaining when it comes to the range of diversity needed to develop improved varieties of crops.[3] We feed each other.

The loss of agrobiodiversity—the reduction of the diversity that's woven into every single strand of the complex web that makes food and agriculture possible—has resulted in a food pyramid with a point as fine as Seattle's Space Needle, making it harder and less pleasurable for us to feed ourselves.

I know it feels counterintuitive to contemplate loss, particularly against the backdrop of floor-to-ceiling aisles in supersized supermarkets. In a Walmart (the number one grocery chain in America[4]) in Winston-Salem, North Carolina, I counted 153 different flavors of ice cream and eight different brands of yogurt. But then I looked further. The choices are superficial—primarily in flavor and secondarily in brand, most of which are owned by the same company. In addition, more than 90 percent of every container of yogurt, milk and ice cream is made with milk from *one* breed of cow, the Holstein-Friesian, known as the highest-producing dairy animal in the world.

I counted 21 kinds of potato chips, but in the produce aisle, I found only five types of potatoes. Most of the bagged potatoes didn't include names, only colors—red, white, yellow—plus "Idaho." The orange sweet potatoes were loose and stacked high. Now I understand why most of those chips I saw just listed "potatoes" as their primary ingredient. Despite being the top vegetable consumed in America, potatoes have been relegated to the background, the carrier for vinegar and salt, sour cream and chives.

Bananas—America's most popular fruit—also carried only a single descriptor: "banana." Although no variety was listed, I knew it was the threatened Cavendish. There are over 1,000 varieties of bananas grown in the world; however, the one that ends up on supermarket shelves isn't the one that has the best texture or taste, but is one that transports easily and has, so far, managed to beat back disease.

I saw six kinds of apples, including Granny Smith, Gala, Fuji and the mealiest, most inappropriately named apple: Red Delicious, one bred for beauty, not taste. Apples were among the first fruits to be cultivated. The original was likely small and tart, closer to what we think of as a crab apple. But, through breeding, we slowly transformed its texture, taste, color, size and level of sweetness. There are now 7,500 varieties of apples grown all over the world, less than 100 of which are grown commercially in the United States.[5] In fact, nearly every historic fruit and vegetable variety once found in the United States has disappeared.[6]

For millennia, we've made decisions about what to grow or not grow—and what to eat or not eat. That's what agriculture is: a series of decisions we, and our ancestors, have made about what we want our food and food system to look and taste like. But our ability to make these decisions—and indulge in our pleasures—is being compromised in ways that are unprecedented.

While some places in the world are experiencing an increase of diversity in certain parts of their diet, the general trend is the same one we see in phones and fashion: standardization. Every place looks and tastes more similar—and the country that sets this trend is America. The refined carbohydrates, animal proteins and added fats and sugars that make up the majority of our diets have also become the template diet for the world.

This increase in sameness is what conservationist Colin Khoury and co-authors of the most comprehensive study to date on the diversity (and lack thereof) of our food supply call our "global standard diet."[7] The researchers analyzed 50 years of data on major crops eaten by 98 percent of the population. They found diets around the world have expanded in terms of amount, calories, fat and protein, with the greatest number of our calories now coming from energy-dense foods such as wheat and potatoes. In areas facing food insecurity, this is a very good thing.

The researchers also learned that agrobiodiversity within our dietary staples has *increased*. Another good thing. And it makes sense: With globalization, foods zoom all over the world, which explains the popularity of mangoes and the random appearance of lychees in, say, Lawrence, Kansas. In Vietnam, 80 percent of calories from plants used to come from rice; now corn, sugar and wheat have risen in importance, and calories from rice have dropped to 65 percent. In Colombia, palm oil used to be nonexistent.[8] Now nearly half of Colombians' plant-based fat comes from palm, and the country is the third largest producer of palm oil in the world.

But this availability obscures the more challenging truth that Colin and his colleagues discovered: Globally, foods have become more alike and *less* diverse. As the amount of food around the world has shrunk to just a handful of crops, regional and local crops have become scarce or disappeared altogether. Wheat, rice and corn, plus palm oil and soybeans, are what we *all* eat now—the same type and the same amount.

Yes, this increase in carbs, fats and proteins has helped feed hungry people, but on a global scale it's also increased our chances of becoming what author Raj Patel calls "stuffed and starved."[9] The world overconsumes energy-dense foods but eats fewer foods rich in micronutrients (the small but essential amounts of vitamins and minerals we need for healthy metabolism, growth and physical development). While 795 million people go hungry, over 2 billion people are overweight or obese.[10] And both groups suffer from micronutrient malnutrition.

The global standard diet is changing the biodiversity of nearly every ecosystem, including the 100 trillion bacteria that live in our gut, part of what's known as our microbiome.[11] The foods and drinks we consume add to or, increasingly, *detract from* the diversity of our intestinal flora and have implications for how healthy or unhealthy we are over the long term.[12]

The factors that contribute to this change are complex and interconnected, but the main reason for this shift is that we've replaced the diversity of foods we used to eat with monodiets of megacrops, funneling our resources and energy into the cultivation of megafields of cereals, soy and palm oil. As farmers from all over the world move toward growing genetically uniform, high-yielding crops, local varieties have dwindled or disappeared altogether. This is why we are now facing one of the most radical shifts we have ever seen in what and how we eat—and in what we'll have the ability to eat in the future.

According to the Food and Agriculture Organization of the United Nations (FAO), 95 percent of the world's calories now come from 30 species.[13] Of 30,000 edible plant species, we cultivate about 150.[14] And of the more than 30 birds and mammals we've domesticated for food, only 14 animals provide 90 percent of the food we get from livestock.[15] The loss is staggering: Three-fourths of the world's food comes from just 12 plants and five animal species.[16]

While these numbers are rough estimates, they speak to a startling trend: We rely on fewer species and varieties for food and drink—a treacherous way to sustain what we need in order to survive. It's dangerous for the same reason investment experts tell us to diversify our financial holdings: Putting all our eggs in one basket (either figuratively or literally) increases risk.

A reduction in agrobiodiversity places us in an increasingly vulnerable position, where warming temperatures or a single pest or disease could severely compromise what we grow, raise and eat. This was, in part, the cause of the Irish potato famine of the 1840s, when one-third of the population was dependent on potatoes for food and one-eighth of the population (about 1 million people) died when a disease known as potato blight ravaged the crop. It also contributed to Southern corn leaf blight, which wiped out one-fourth of American corn in 1970. And now it exacerbates the proliferation of wheat rust, known as the "polio of agriculture," which is threatening 90 percent of African wheat.[17]

It's why plant geneticists are working around the clock to develop a new type of banana to replace the Cavendish, a variety that was introduced when the soil fungus *Fusarium oxysporum*, in the 1950s, wiped out the Gros Michel—the banana that *used* to be the one on store shelves. Those Cavendishes are now succumbing to Tropical Race 4, a strain of the same fungus that decimated the Gros Michel.[18]

The depletion of agrobiodiversity also includes what scientists call "genetic erosion." Stefano Padulosi, senior scientist at the conservation research institute Bioversity International, explained to me—in another 20-minute interview that stretched to two hours—that the erosion manifests in different ways; some of the changes we see and some we don't.

Stefano is a plant explorer known for his work on finding and saving neglected species of foods, ranging from pomegranates to arugula (the latter of which has earned him the nickname "Rocket Man"). I met him to better understand how the industrialization of seeds has transformed what ends up on our plates. What I learned was that industrialization is just one of *many* reasons for our limited food choices and changing diets.

"When an entire set of traits that make a certain variety or breed distinct from another is lost altogether," Stefano said, "then we talk of the loss of that variety or breed. From a scientific point of view, variety is defined as a good combination of traits—like adaptation, taste or yield—but variety is also an expression of terroir, food culture and identity of people." In other words, erosion is both genetic and cultural. These losses—and, in some places, slight gains—are due to a wide range of social and environmental reasons: from how we manage our land and financial markets to changes in where we live and what we eat. (A comprehensive list written with the help of Stefano, Luigi and FAO's Paul Boettcher is included in Appendix II.)

Take, for example, the pistachio. When Stefano told me about the tiny nut, I finally understood how invisible a lot of this genetic and cultural erosion is—and how dramatically our diets have changed for reasons that don't immediately connect back to food. The transformation of the pistachio industry was the unintended consequence of political strife, part of a cascade effect of trade restrictions that were meant to punish the captors of hostages. It had nothing to do with food or farmers.

Iran used to be the center of the world's pistachio industry. Those little green nuts are actually seeds that Persians bred to split open, and they come from the same family of plants (Anacardiaceae) as mangoes, cashews and poison ivy. An integral part of Middle Eastern foods and celebrations, pistachios originated in Afghanistan and are one of Iran's biggest exports after petroleum. Evidence of the nuts dating back to 6 BC has been found in both of these countries.[19]

In 1929, botanist William E. Whitehouse traveled to Persia (now Iran) to collect pistachios in hopes of finding a variety that would be suitable for growing in America. Of the 20 pounds of nuts he gathered, only one variety flourished—in California's San Joaquin Valley.[20] To put this in perspective, a single nut weighs one-fortieth of one ounce. There are 320 ounces in 20 pounds. Out of everything he collected, *one* nut (seed) took root.

Food is bound to place. That small female nut was, at that time, the only one that could handle the climate and other environmental conditions of the United States. Whitehouse named the pistachio "Kerman," after a famous carpet-making city near the birthplace of the nut.[21] The tiny but mighty Kerman built a fledgling American pistachio industry that started to blossom in the 1960s and exploded decades later when, in 1980, President Jimmy Carter instituted a full trade embargo on Iran as a result of the 444-day hostage crisis. This included all agricultural products.[22]

The ban devastated the Iranian pistachio market and empowered the United States to build its capacity for pistachio cultivation. Today, America is one of the world leaders in its production. The nearly 520 million pounds of pistachios that were grown domestically in 2014 descended from that one Kerman, a variety that represents almost all of what is planted.[23]

When Stefano and Luigi first told me about the reduction in agricultural biodiversity, I was incredulous. I had come to Rome to do research on seeds yet knew nothing of what they described. I had spent my life obsessed with food—and it was disappearing? Why hadn't I heard about this? How was this possible? The answer lies in the fact that many of these changes have happened slowly, over time. These losses in food are buried in the soil, tucked in beehives and hidden in cattle feedlots. They start with microorganisms invisible to the naked eye and echo through every link in our food chain—from soil to seed to pollinator, from plant to fish to animal—compromising the very ecosystems that make much of our food possible.

The loss of agrobiodiversity has and will transform not only what and how we eat but who will have the resources to eat at all. Because behind every one of these foods and drinks are the people who rely on them for their livelihoods—from field hands and factory workers to grocery clerks and chefs.

How do we feed one another? And how *will* we feed one another? It's impossible to escape the headlines and news reports expressing concern about food security for our growing population. But what I care about is feeding myself and the 805 million people who are hungry *today*. This includes more than one in five children in America who are food insecure. A statistic that, when broken down by race and ethnicity, becomes even more heartbreaking: Almost 40 percent of African-American kids and 30 percent of Latino-American children are undernourished.[24]

For the past 20 years, the rate of global food production has increased faster than the rate of global population growth.[25] The world produces more than one and a half times enough food to feed everyone on the planet, which is also enough to feed the population of 9.6 billion we anticipate by 2050.[26]

This matters because a lot of the changes we see in food and agriculture have been made in the name of feeding hungry people. But the challenge isn't simply an issue of availability; it's one of access. Food and the resources required to buy food aren't efficiently or equally distributed. That's why the hungriest people in the world are smallholder farmers—the over 500 million people responsible for feeding the majority of the world's population.[27]

The people who grow food are too poor to buy it.

The majority of these farmers are women, most of whom live in extreme poverty, which is why they are moving to cities and entering the formal workforce in higher numbers.[28] (Women have always worked, but they aren't always recognized or paid for it.) Add to this fewer home gardens, less time to grow food, the exponential growth of supermarkets and fast food joints and a bit more money with which to buy cheap, processed food. You've now got a recipe for the global standard diet.

I grew up in a household where my mom cooked dinner almost every night. Most of our home-cooked meals consisted of rice with peas (I hate peas) plus some curry or *dal* (lentils), with salad and fruit for dessert. The food I ate outside of those confines was completely different. I'm from the South, where sweet tea is the beverage of choice and grits and barbecue are dietary staples. From an early age, my tongue knew the sting of chilies and the bite of masala alongside the saltiness of a perfectly baked biscuit and the cloying, but heavenly, sweetness of pecan pie. Even today, my comfort foods are *bhartha* (roasted eggplant) and rice (sans peas), and macaroni and cheese and collard greens.

Every person who has immigrated to the United States has a similar mash-up and story about how food was a source of solace—but also one of shame. There were many days when I didn't want to come home to a kitchen reeking of fried onions, ginger and garlic; I wanted TV dinners, meatloaf, whatever would allow me to fit in rather than stand out. But I now realize these tastes are not only the essence of who I am but what America is: a melting pot of cultures and flavors ranging from soul food to sushi.

Yet, for many cultures, the stigma of difference has magnified and deepened. It's what Braulio Ferreira de Souza Dias, head of the Convention on Biological Diversity, explains is another contributing factor to the loss of agrobiodiversity: the replacement of traditional foods deemed to have "low status" with processed foods that seem more modern. If my mother had stopped cooking Indian food, it would have become an exotic novelty, not a mainstay. I wouldn't have a deep appreciation of those foods—my foods. I wouldn't miss them if they were gone.

When I was 26, I moved to India. My first year there, I feasted on all the Indian foods I had missed while growing up in the United States: *dhansak, bhel puri, pav bhaji*. During my second and final year, the trend reversed itself, and I started to miss another set of tastes. I stockpiled arugula for salads (on the rare occasion my vegetable seller had it) and made lasagna with noodles that perpetually went limp. The meals were close-not-quite approximations to what I knew and loved. When McDonald's opened up in my neighborhood of Bandra, I was secretly thrilled. I'd finally get decent fries and a chocolate shake. McDonald's was also a taste of home.

Nearly 20 years later, those fries and shakes are no longer a novelty, found in 300 outlets in India, part of over 34,000 McDonald's in 116 countries around the world.[29] They are so ubiquitous that Big Macs are used by economists as an informal way to determine if currencies are over- or undervalued.[30] Cheap, processed food has changed the world.

Fast food doesn't just provide momentary gustatory relief for someone longing for a taste of the familiar; it has now reshaped what taste means for *everyone*. Chain restaurants go to great lengths to make a burger in Poughkeepsie taste the same as one in Paris, with a few concessions for what the company calls "locally relevant ingredients."[31] From

Austria to Vietnam, the Golden Arches and countless other fast food outlets have changed the shape, size, taste and speed of food.

This is because many countries are eager to emulate what America stands for and to avail of freedoms that appear to make life more convenient. They want the right to enjoy what the rest of the world eats—or at least what the world *says* they should eat. The loss of agricultural biodiversity and our narrowing global diet are, in part, a reflection of hamburger culture. Fast, processed food is considered modern, easy and cool—sentiments that wind all the way back to the crops that are grown.

Phrang Roy, coordinator of the Indigenous Partnership for Agrobiodiversity and Food Sovereignty, saw this surface in the 1960s during the rise of industrialized agriculture, which started with the Green Revolution, a period of intensive planting of high-yielding cereal crops that transformed agriculture. "If a farmer grew a multitude of crops, he was not modern, he was backward," Phrang said. "If his field was a bit messy, not manicured or monoculture, he was backward. This is the psychological impact of accepting someone else's standard. He feels shame for not using chemicals or high technology."

This standard has continued and extends from the field to the plate. The idea of "good" food has been reduced to foods that grow abundantly and stand strong in the field—and can sustain long journeys and continue to stand strong on supermarket shelves. And while these qualities are important, they give short shrift to the many reasons we choose one food over another, regardless of our budgets.

They also inform the steady refrain from various players in the food industry that the reason we should settle for inexpensive, marginally nutritious food is because we can't afford anything better. Some of the nonprofits, government officials and businesses who maintain these views mean well: Hunger is real and people need to eat. But others, especially those profiting from cheap food, use poverty as an excuse to keep feeding us food of lower quality, sowing fields full of nutrient-poor monocultures and glutting shelves with processed foods chock-full of an addictive combination of salt, sweeteners and fats that keep us coming back for more.[32]

We all know this craving. The kind of yearning that has some of us driving to the grocery store at odd hours of the night—or lining up for fries at a McDonald's in India. Processed foods are engineered for hyperpalatability, a kind of uber-deliciousness that comes from a combination of fat, sugar, salt and flavorings that can only be cooked up in a laboratory. They ignite the same brain circuitry that fires when we're addicted to a drug and want another hit. A medium serving of McDonald's fries contains 19 grams of fat, 270 milligrams of salt and 19 ingredients.[33] Contrast that with a potato: less than one gram of fat, 13 milligrams of salt and one ingredient. Hyperpalatable foods are like crack cocaine for our taste buds.[34]

This addiction can be easily confused with love. But eating what our body *truly* loves requires us to pay attention to how our system responds to what our head says it wants. The synthetic trans fats, high-fructose corn syrup and artificial sweeteners, colors and preservatives in cheap food make us feel good in the moment but are the reason that, for the first time in over two centuries, the current generation of children is predicted to have a shorter life expectancy than their parents.[35]

The refrain of affordability also obscures deeper challenges around why people aren't paid enough to spend more money on food. Why is this all we can afford? And why isn't food a bigger priority? We Americans spend less of our income on food—6.7 percent, with an increasing percentage spent on snacks—than almost any other country on the planet.[36] Less than what we spent on food during the Great Depression.[37]

No one should be underfed. Those who are should have greater support in not only being fed but being fed *well*. And those of us who can afford more might do well to reconsider why we pride ourselves on cheap food. That bargain means that someone (a farmer or factory worker, not the CEO) or something (such as safety or environmental standards) was likely compromised. "The best price isn't necessarily the cheapest one," explains Dr. Dias, "because what is cheapest is usually the lowest quality."

I was sitting at my kitchen table in my tiny apartment in Rome when Dr. Dias said these words to me during our short phone interview. As I held my mobile to my ear, I picked at the remnants of lunch

I had pushed to the side to make room for my interview notebook. My meal consisted of fresh figs, a few thin slices of prosciutto and pizza bianca (thin, salty pizza dough absent of any toppings that's easily found in Italian pizzerias). The lunch had cost a few euros, roughly the equivalent of a Happy Meal. But the figs were in season and the ham came from a butcher I had befriended. The humble meal was delicious and represented the best that the season and region had to offer. Cheapest is usually the lowest quality—but it doesn't have to be. We still have a choice.

"What is often the case," Colin Khoury told me, "is that the bigger the system that grows or manufactures the food, the less of a tendency there is toward diversity. The scale of production favors sameness." Our industrialized food system was designed for efficiency and yield—not nutrition, taste or diversity.

"Diversity isn't a finite thing," anthropologist Pablo Eyzaguirre clarified during our interview in his plant-filled office at Bioversity International. "It's a dynamic construct that people create all the time." And it's not the same in every place. As we migrate from place to place, our foods travel with us. "They create new diversity as people look at them in different ways. Take the cowpea, what's known in the United States as the black-eyed pea. In most parts of Africa, the leaves are the primary food. When you're selecting the crop for leaves rather than beans, you're envisioning a different kind of plant. Cultures see plants in different ways. That's why it's so important to maintain diversity in what we grow and in who grows it. Agrobiodiversity gives us options." He added, "When you take the culture out of it, its use becomes very limited, with implications for both the gene pool and the potential of the species." There are also implications for taste, which is where my appreciation had taken hold.

Soon after I started researching this book, I saw a bumper sticker that read "Extinct is Forever." It's true. It's what we face every time we shrink agrobiodiversity from thousands of varieties down to a handful. We stop growing it, we stop eating it and, slowly, it disappears. The loss of genetics is accompanied by the loss of knowledge on how to grow foods and how to prepare and eat them. It's the cultural erosion that accompanies the genetic one: Our culinary traditions are going extinct, too.

No matter where you live, you have the memory of something you used to eat that is no longer a part of your diet—something your grandmother used to make, something a small shop used to carry. Something you have lost. This extinction is a process; it happens one meal at a time.

Fortunately, a lot of these changes have occurred in the last few decades, which means they can change again. That is, of course, as long as we sustain the diversity found in the wild, on farms and in stored collections that contain the traits we might need now or in the future: immunity to a disease, greater adaptation to a changing climate, the possibility of higher yields or greater nutritional value—and delicious taste.

But in order to support this diversity and facilitate change, we have to start thinking differently about the food in our fields and on our plates, and be more discriminating about its sources. "How do we buck the system just a little bit?" Colin asks. "Think of oil. We're definitely eating more of it: soybean oil, then palm oil—much more than other oils around the world. Although it isn't immediately obvious that eating olive oil would be radical, in the big picture that's exactly what it is. Eating olive oil is now a radical act. Eating anything that's not rice, wheat, corn, soy or palm oil is radical."

The revolution starts here, on our plates, by looking at the pillars of our own diets and by making simple changes. The way to take back this power for ourselves is to understand why we eat what we eat. And to understand what we're losing—so we know what to reclaim.

II. TASTE

Tasting is different from drinking or eating. Eating is an act of digestion: lift substance to mouth, chew, swallow, process. *Tasting* is something else. It requires us to slow down, pay attention and savor—and demands everything we've got: ears, eyes, nose, mouth, head, heart. Tasting is about getting intimate with the substance we have actively chosen to put inside our bodies—the beer that makes our tongues tingle, the chocolate that melts in our mouths. It happens in the immediacy of the moment but, simultaneously, reflects the long history of who we are, as well as the flavors of our collective memory.

Understanding the motivations around why we eat what we eat, or don't eat, is a critical first step in reshaping our food and food system. It feels like a simple act—find food, put in mouth, eat. It is, and it isn't. Food, sociologists explain, is created within our cultures and consumed and shared in every space we occupy: home, work, church, school and beyond. Our acts of consumption are woven into every part of our existence, from births to funerals. They define who we are. Every minute of every hour of every day, someone is eating; millions upon millions are eating.

"We are what we eat, and we eat what we are," Jerome, my talented chef-artist friend, announced during our trip to Morocco two years ago (on one of my few breaks from book research). I rolled my eyes and chalked it up to his eccentric nature, but he was spot-on. We filter taste not only through our personal experiences but also through communal ones. Taste is a reflection of who we are, what we've been exposed to and what's expected of us in a given group or society. Eating with your hands, for example, is expected in Ethiopia but impolite in England. Horses and dogs are food in one community and friends in another. And just because we know something is healthy doesn't mean we're going to eat it. It's not just what we eat but also what we *won't* that defines us.

Taste is so multifaceted that it's sometimes hard to understand why we love what we love (or hate what we hate). Of course there's deliciousness. But even that's shaped by forces outside of us, including the predispositions we come into the world with (informed by human evolution and our mother's taste preferences), plus what we cultivate once we get here.

First, there are physiological reasons, starting with the nose and mouth. Taste is predominantly about aroma—what we smell—and secondarily about what we experience in our mouths. These senses are distinct but culminate into one experience. We commonly use the word "taste" to describe this coming together, but what we actually mean is "flavor"—the moment when what we put in our mouths and what we smell converge. (There isn't a verb for the smell-taste mash-up, so I use "taste" throughout the book but make the distinction whenever possible.) Odor perception expert Johannes Frasnelli summarizes it best: "We perceive the flavor of food via the sense of smell."[38]

That's what happened when I sat in Camino and lifted the glass of wine Ali had brought me to my nose and sniffed. Everything that has a scent gives off volatile (airborne) molecules that float through the air and into the nose and mouth. They trigger smell receptors in the cells of the nostrils and in the airshaft that connects the nose to the mouth, otherwise known as the retronasal passage. These aromas stimulate a small patch of tissue called the olfactory epithelium, located high in the nasal cavity. The smell receptors within the olfactory epithelium catch those odors and send messages on to the brain: *This wine smells terrific.* The signals then get transmitted to multiple parts of the brain, many of which are part of the limbic system—a set of structures responsible for our emotional responses, including our formation of memories: *This wine reminds me of rolling down hills of grass when I was five.* That's why smells can take us all the way back to our childhood in ways that other senses can't.[39] "Hit a tripwire of smell," Diane Ackerman writes in her stunning book *A Natural History of the Senses,* "and memories explode all at once."[40]

Smell is fundamental to our understanding of flavor, a sensation that concentrates in our mouths where taste receptors help us distinguish five basic qualities: sweet, sour, bitter, salty and umami, plus the presence of fat. *Umami* is a Japanese word that means "pleasant savory taste" and is used to describe the meaty, earthy taste we find in foods like fish, meat, mushrooms and aged cheese (and in our first food, breast milk).[41]

But the experience of flavor isn't limited to the nose and mouth; it's also something we do with our eyes. We're drawn to foods that look enticing, and we perceive them differently based on their color, shape, size, texture and packaging. Our experiences of foods and drinks are also influenced by cutlery and serving dishes, as well as the environments in which we consume them. A candlelit dinner with a beloved, for example, tastes more delicious than that same meal eaten as leftovers, reheated in the office microwave and consumed under the glow of fluorescent lights. Context matters.

In fact, a number of studies have shown certain glass shapes optimize our perceptions of aroma in wine, and candy bars with a rounder shape are perceived as sweeter than ones with more angular shapes (which are considered bitter).[42] Russ Jones, creative director at the sensory strategy firm Condiment Junkie, explained their comparable study of hot chocolate: The same drink served in different colored mugs, he said, drew out completely different responses. People felt hot chocolate served in red mugs tasted richer, while chocolate in blue mugs tasted milder and, in black, tasted bitter. "When we asked people to give a reason, they all said something about the chocolate, but that was the same. It was all because of sensory cues. Color completely changed their perception."

Sound also impacts how we perceive what we eat and drink: A crisp snap or bubbly fizz can be an indication of freshness or quality. It's also likely, based on initial research from Condiment Junkie, that sound amplifies the experience of what we consume.[43] "We're always cross-linking our senses," Russ told me. "Listening to crashing waves and seagulls ... makes fish taste fresher and fishier, because it reminds you of the freshest fish you ever had by the sea. French food accompanied by French music, Indian food accompanied by a sitar ... It intensifies the experience."

We prefer foods that meet our sensory expectations, including our sense of touch. Depending on the food, this sensation is described as "body" or "mouthfeel." It's the gooeyness of a grilled cheese sandwich, the sliminess of okra, the way bacon crunches in our mouth and coats our tongue with grease. We want these expectations to align.

I had a vague idea of how smell and taste worked, but before I started writing this book, I hadn't given much thought to the full experience of our five senses, or to external influences and the ways in which we carry our life story to our plate. And yet, I knew all were at work.

During my Moroccan adventure with Jerome, following a lunch of dried fava bean soup and steamed camel hump, I started weeping on a street in Fez. Not crying, *weeping*. It happened immediately after taking my first bite of *sellou*, a rich sweet made of sesame seeds, almonds and flour seasoned with anise seed that's traditionally served during Ramadan or after childbirth.

Jerome tried to console me, but he didn't know what was wrong. My tears had nothing to do with the pastry, place—or ingestion of camel. What the *sellou* triggered in me was the taste of something my grandmother used to make for me: *pinnis* (pronounced "pin-neez")— sweet balls of chickpea flour, raisins, cardamom, sugar, ground flaxseed and almonds. They are what Indians consider "heaty," a food that warms you up and gives you energy during cold winter months. Because they're labor-intensive and take multiple days to prepare, most people eat them in small quantities. Not I. My grandmother would spend weeks preparing a big tin of *pinnis* in anticipation of my arrival in India. I gobbled them down, never conceiving of a time when I wouldn't be able to eat them.

My nani died in August of 2007. And then, six years later, there she was. Not just the taste of something she cooked but the memory of her. Her being, her *presence*. That bite of *sellou*—that edible memory—brought my nani to me, right there in the Fez medina.

French sociologist Pierre Bourdieu asserted that culinary taste is a reflection of multiple aspects of who we are. This includes our education level and socioeconomic status, which inform what we can afford to eat in terms of both money and time, and the effort that goes into making food. Taste is also influenced by experiences with our families and communities, and our comfort with the hierarchies and behaviors that are considered appropriate for our social groups (known as social norms). The use of chopsticks, for example, might take on a completely different meaning depending on whether we're from Chicago or Shanghai.[44]

The entire experience of flavor exists within these cultural contexts and affects our entire sensory experience. How did we learn, for example, what sour tasted like? Was it from a grapefruit or a gooseberry? Does our point of reference for sweetness tend toward sugar or honey? When we think of fruit, do we conjure apples or mangoes? These frames of reference are reflected in everything that follows, and are constantly evolving, both personally and culturally.

Take lobster. The crustaceans are bottom feeders, the oceanic equivalent of rats, or what my friend's mother used to delight in calling "cockroaches of the sea" just before serving them to us. Before the mid-19th century, they were what fishing communities in the northeastern United States ate only when they had run out of the good stuff. Lobster was used as fish bait and served as prison fare. Now those cockroaches of the sea are a mark of sophistication.[45]

Bourdieu argued taste isn't a pure construct but a strategic tool born out of a system where our likes and dislikes become a way to distinguish ourselves from others (*I start my day with a Diet Coke* versus *I start my day with green tea*) and connect us to our lineage (think of halal or kosher dietary laws). In its best manifestation, these distinctions can be a source of pride; in its worst, a form of judgment. It's not enough for something like lobster or kale to exist. We need to have *access* to it.

This is one of the biggest criticisms of so-called food deserts: areas that experience a dearth of fresh, nutritious foods. We need to be able to afford them, but, more importantly, we need to feel like they belong to us—that they have a rightful place in our communities and on our plates.

In my youth, kale did not belong to me. Now it does. To be clear, I don't really like kale, but I eat it for its virtuosity. It's a nutritional powerhouse.[46] Until I knew this, I wouldn't have chosen it. Ever. Kale isn't a food I grew up with, and I have no emotional connection to it. But now that I know how good it is for me, I'm cultivating an appetite for the giant bumpy leaves that were bred from members of *Brassica oleracea*, the wild cabbage family that includes Brussels sprouts, broccoli and cauliflower. Kale paves my food path—one that later leads to chocolate—with nutritional vigor and virtue.

Our identity and community are shaped and revealed through everything we consume. We are what we eat, and we eat what we are.

This is what we bring to the table every time we sit down: our wants, needs and values, and the stories of who we are—our ethnicity, gender, geography, history, class and worldview. We are also influenced by a host of factors that exist outside of us: trade agreements and restrictions, legitimate scientific discoveries and persuasive marketing ploys, planetary shifts and cultural trends. It's limiting and inspiring. It's what I brought to the moment when I turned to that lovely server Ali and said, "Surprise me."

Taste is something we all do and all have. It doesn't just belong to foodies or sophisticates. Each of us owns and shapes this construct. What we feel about tasting and eating—what we savor—shouldn't be discriminatory or hierarchal. Because if we operate from the premise that only certain people own taste, then there is no point in exploring. We should all just frequent the places that have collected the most promising Yelp reviews or greatest number of Michelin stars.

I refuse to do that. I refuse to let someone else define what delicious is for me. To whatever extent I can, I want to define and redefine what tastes good. I want to define what *is* good—for me. And I want you to do the same for you, because taste is both universal and personal. What each and every one of us cherishes matters.

I have found deliciousness in a bucket of yucca (accompanied by large plastic cups of rum) on a farm outside of Havana; in perfectly fried eggplant served in a Mumbai *dhaba* (diner), where my girlfriend and I kept our feet hovering just above the floor to avoid contact with cockroaches; and in my aunt Toshi's kitchen. Great tastes are *everywhere*. Sometimes they're fancy, but most of the time they are not.

Finding those tastes requires less of an open wallet and more of an open mind and heart. It's how I learned I actually enjoy curried horse flank but am not crazy about caramelized mealworms. I gave them a shot. And I would be willing to do it again. We can only truly know our preferences if we dare to explore what's out there. (As possible incentive, researchers have found that those who eat a varied and diverse diet— what they categorized as "adventurous eaters"—weighed less and might be healthier than those with more conservative diets.[47])

This isn't easy because we're hardwired to resist change. As kids, most of us display what researchers call food neophobia, a fear or rejection of new foods. The two ways we transcend these fears are by learned safety (*I ate the new food and it didn't kill me*) and food exceptionalism (*I don't like new foods, but this is an exception*). In the case of exceptionalism, researchers have found the best way to appreciate new foods is through direct experience.[48] In other words, the best way to reconcile hesitation or fear about eating new foods ... is by eating new foods. It's like telling someone you love them. Being vulnerable and taking that step into the unknown is the only way to know what's out there. And the process is iterative: Our threshold of acceptance—our love—grows every time we take the risk.[49]

Risk, however big or small, requires a kind of courage that, through this journey, I now understand can be mapped onto our lives as a whole. "You are here to risk your heart," writes Ojibwe author Louise Erdrich in her masterful work *The Painted Drum: A Novel* (P.S.). "You are here to be swallowed up. And when it happens that you are broken, or betrayed, or left, or hurt, or death brushes near, let yourself sit by an apple tree and listen to the apples falling all around you in heaps, wasting their sweetness. Tell yourself you tasted as many as you could."[50]

These words and tastes have become my touchstone, as I try to savor all I can, both bitter and sweet. Every new bite, every new sip, can change us—if we are present and open to it. In that moment of saying yes to Ali, I didn't know this. I did not expect that glass of wine to transform me. But occasionally a taste reaches out and finds us. It wasn't what I knew wine to be. I took another sip. And another, and another, and returned every evening for the next week to, again, taste that wine and experience that story. Wine was only the beginning; I wanted to taste it all.

This book is the product of years of grief, frustration and longing. Grief over the realization that the foundations of food and food itself—the most delicious, diverse varieties of food—are being lost slowly and irrevocably. What we do to food, we do to ourselves. When we all eat the same variety of apple or banana, we aren't just losing genetic diversity; we're losing part of what makes us who we are. Frustration that there seemed to be little I could do about it. I didn't understand the myriad reasons for this shift or the deep origins of the foods at risk of changing. I didn't know how to bring what seemed so far away (farmers in fields and workers in factories) up close. I didn't realize how much my sense of well-being and pleasure was bound up with theirs. Or how we, the eaters, could be a part of the solution.

Until now.

This book is for anyone who has bitten into a grocery store tomato in winter and longed for something more. For anyone who has settled for blandness but yearned for something deeply satisfying. The foods I selected—bread, wine, coffee, chocolate and beer—feed me in different ways. One wakes me up, another brings me down; one nourishes my heart, and the other tends to my soul. What ties them together is that they are my constants, woven into the fabric of my life.

Through these five foods, and our five senses, we'll look at stories of loss and interconnection, exploring what is disappearing and why, and learning how we can save agrobiodiversity through various forms of conservation and through the choices we make, with every sip and bite we take.

But it's not enough to be told to pay closer attention or eat differently if we're not given the tools to make the change. That's why each chapter includes an investigation of our senses (smell, taste, touch, sight and sound), an explanation of how they shape our experience of eating and drinking, and a tasting guide (in the final section of every chapter) that helps turn that awareness into a delicious, tangible experience.

Taste is the gateway through which we will transform food. By savoring foods like we never have before—by demanding what's delicious—we can transform what is grown and sold. It is the first step in reclaiming what we love.

WINE

I. IN THE GLASS

Tasting wine felt out of my league. It was fancy; a beverage that I had, on more than one occasion, swirled out of the glass and, only once, accidentally sniffed up my nose (a Merlot, not pretty). Wine was, for me, the same as it is for many: the social lubricant that made holiday parties easier, an intoxicant that made the candlelit corner of a restaurant a bit more romantic, the substance that smoothed the rough edges off a day. But it wasn't something I had any intention of exploring.

My God, I had been missing out.

As part of his 1941 research on the nature of pleasure, German psychologist Karl Duncker asked "whether the object of pleasure is the wine, the drinking of the wine, or the sensory experience of drinking the wine." In the case of wine, he said, "The answer is clearly the last, the experience of the flavor."[1]

In addition to inducing pleasure, relieving stress and, possibly, igniting romance, wine (in moderation) slows brain decline and increases longevity.[2] And we have been reaping its benefits for centuries.

In ancient Greece, wine was so highly esteemed that it held a place above food.[3] Greek mythology ranked the gifts of Dionysus—the god of the grape, the harvest and the wine—on a par with Demeter's gift of grain. It made water drinkable and fostered camaraderie, not unlike our modern-day holiday office party. But what set wine apart from other beverages was that, in many cultures, it was also considered a truth serum. *In vino veritas*, the Latin adage goes: In wine [there is] truth. Germans,

Italians and Persians believed the true nature of a person was revealed in the glass.

It turns out one of the truths revealed in *my* glass was insecurity. Not only was I judging the taste of the wine, I was also judging my own taste.

This judgment and self-doubt were unnecessary. Study after study shows we're all vulnerable to the power of persuasion, from fancy labels to verbal cues. No one—neither expert nor amateur—can consistently differentiate between fine and cheap wines or distinguish the diversity of flavors within them. Our appreciation of wine and other drinks and foods is as much a construct as any other sensory experience. Physiological, psychological and cultural—we bring our stories to the glass, along with all the trappings of taste Pierre Bourdieu described.

The most challenging moments in my explorations of wine, coffee, chocolate, beer and bread were the times I lost sight of this. In one course on sensory analysis (the practice of using all our senses to analyze food and other items), the instructor made a game out of identifying scents from Le Nez du Vin—a kit of 54 liquid vials approximating aromas found in vineyards, grapes and wines. Le Nez breaks these smells down into five categories: fruit, floral, vegetal, animal and grilled. It's intended to help people connect with some of the over 400 aroma compounds present in wine.[4]

The game came at the end of the week, the culminating activity of our final class. I was the one person in the room (aside from the teacher) who had actively studied scents, but when it came time to play, I had no interest in entering my nose or retronasal cavity into competition. I was tired and heavyhearted, in the middle of *another* break within a five-year relationship that would eventually end. I was ready to go home.

Vial after vial passed before us. To me, all the scents smelled artificial, and nearly everything I identified was wrong. I smelled candy when everyone else smelled butter; I found detergent where others found car tires (the vial was labeled "rubber"). I couldn't tell the difference between vial 19 (apricot) and vial 20 (peach). I finally just gave up and started saying I didn't know.

"You don't know?" the instructor asked, with what may have been a hint of surprise.

I slowly turned my head from side to side. "I have no idea."

Yes, I hate losing. But the saddest part of the exercise was that it turned something I loved into something arduous. After our sniff test, we sat with the same teacher and tasted chocolate. Every student in the room looked to him for cues, watching tentatively as he screwed his eyes shut and focused inward. "Apricots," he announced. "Honey." I tried to find those same things but couldn't. Not on that day.

So I pretended, which made me feel ashamed. Then I went home and—to rid myself of shame and assuage my fractured heart—inhaled chocolate. I didn't taste it; I scarfed it down. And it felt good.

I share this story to say, don't be as hard on yourself as I was. There is so much that brings us misery and makes us feel inferior. Don't let it be chocolate or wine. Pay attention and be present. Own your choices. Eat and drink with reverence and gusto, whether it's a Big Mac or a mountain of kale. Inhale chocolate when occasionally required—but then return to the slow savor.

I have spent over three years learning from all kinds of food, beverage and sensory experts. Each one shared critical insights, but the most important lesson I learned is one I came to on my own: We can't screw this up. Food is love *and* joy. This distinction is important because love isn't always fun. It is, in fact, often the opposite of fun. But it is enduring. It is what we get out of bed and fight for. It's what we deserve.

I didn't always understand this. When I started writing this book, I had no idea my journey would inspire such profound appreciation for food and drink, for the tireless hands that produce and sustain them, and for the small acts of courage required in asking for what I wanted. It was a lesson that, for me, began with wine.

In wine appreciation, a handful of factors separate the rookies from the pros. Broad exposure is certainly one, but the most important distinguishing feature is having a language and reference for what we experience. Aficionados have not only more contact, but the vocabulary and memory recall to express what they find in the glass. And this isn't limited to wine. Randy Mosher, author of the brilliant book *Tasting Beer: An Insider's Guide to the World's Greatest Drink*, says, "If you take the time to develop an approach and vocabulary, even casually tasted beers may reveal themselves in greater depth, meaning, and eventually, pleasure."[5]

A common language doesn't just foster a deeper connection to the food or drink; it also connects us to one another. For example, when we connect flavors to shared understandings of stone fruit or a favorite spice, we can better appreciate the notes of cherry and cloves in, say, a Napa Valley Zinfandel. Language allows tasting to become more of a united experience, and helps provide a level of consistency in how we evaluate what we eat and drink.

Sensory expert Darin Sukha, a research fellow at the Cocoa Research Centre in Trinidad, explains these words and thoughts are part of our "sensory library." When I spent a few months with Darin studying chocolate, he would encourage me to try new things as a way to build up my library. Sometimes he'd stop mid-sentence in a field or on a crowded street and command me to inhale. Once, he crushed a fresh cacao leaf in one hand and a dried one in another and had me compare the two. Other times, he'd make a point of ordering foods he knew I hadn't tried as a way of reminding me to continue to stretch my brain and palate, and then asked me to describe what I was tasting.

I said yes to everything: soursop and lychee; doubles and *saheena*; macaroni pie, fish pie, chicken pie and potato pie. It was delicious. It gave me a better appreciation not only of what I was eating in the moment, but of every food and drink that followed.

In wine, research shows that when novices are given a list of words that appropriately describe and match different varieties, they—we—become more discerning.[6] That's why sensory chemist Ann Noble and her colleagues spent over three years conceiving the Wine Aroma Wheel, one of the first sensory wheels ever created for foods and beverages (beer and scotch preceded it)—and the first wheel developed for wine.

The Wine Aroma Wheel found on page 344 is based on organizational principles from cognitive psychology where objects (in this case, smells) are grouped together according to perceived similarities. For example, if we were going to group items under a broad category, like clothing, we'd create subcategories, such as shirts and pants. Within each of those subcategories, there would be additional nested subcategories, like jeans and trousers. In the 1970s, the category would have also included bell-bottoms and, in the 1980s, stirrup pants. (These constructs are always changing and are a reflection of the time and culture in which they are created.) Ann and her team created the same organization of categories and subcategories for wine aromas.

As I tried to find my words in wine, this wheel became my compass, so I contacted Ann to schedule an interview about the motivations behind its creation. In preparation, I stuffed my head full of chemical analyses and neurophysiology. I read dizzying studies on attitudinal models that showed the distinction between internal and external taste divides. "In discursive terms," one read, "the distinction is understood rhetorically, allowing speakers to construct or suggest the source of a taste evaluation themselves."[7]

Right.

Fortunately, I had a partner in this inquiry: Caleb Taft, then wine director at Camino, the restaurant where I drank my transformative glass of Trousseau Gris. A 42-year-old self-described "weird guy who spends a lot of money on food and wine," Caleb has double-pierced ears, a long blondish soul patch and icy blue-green eyes that hold your gaze. He blends seamlessly into the hip, urban environment that is Oakland. And while his Ivy League degree, extensive wine knowledge and cool demeanor afford him ample opportunity to be pretentious, he isn't. Ever. Caleb is gentle, wise and incredibly patient—especially with me.

The first time we met was in a bar in downtown Oakland, where I shared with him the unexpected joy I got from the Trousseau Gris he'd placed on the wine menu, a wine made by Scott Schultz of Jolie-Laide Wines. I proposed a journey I hoped to take, tracing the grape back to the place where the original plant material came from, then on to the vineyard and the winery. "With you, the man who brought the wine to me." I tried to sound casual, but I really wanted him to say yes.

He said yes.

A few months after that first meeting, Caleb and I headed to the Russian River Valley in Sonoma County, California, to see the vineyard, the place where the grapes had been grown. But first, we had to visit

Ann Noble in order to better understand the foundation of flavor. I reasoned I could show my wine naïveté to Caleb; he meets people who are new to wine every day. But to the University of California, Davis, emeritus professor of enology and creator of the foundational sensory wheel for wine? I had to step up my game. I showed up at Ann's house with Caleb and two single-spaced pages of meticulous questions. I took a deep breath and, at the exact minute of our scheduled interview, rang the doorbell. No answer. I rang it again. Ann opened the door and, with her mouth full of food, asked, "Who are you?"

My eyes widened and I started stammering about the book and the wheel and my journey with the Trousseau Gris.

"Can you come back in a few hours?" she asked.

"Unfortunately not," I said. I explained that we had to meet Peter, the winegrower, and Scott, the winemaker, and then spiraled into a dizzying description of biodiversity and my pending journey to Ethiopia to learn about coffee and to Ecuador to learn about chocolate and, finally, I just trailed off into a "Would you please ... ?"

She took a minute to size us up and then invited Caleb and me into the kitchen to join her for lunch. Seated around her wooden dining table were four guests. They were drinking wine and eating olives, cheese, bread and salad greens from Ann's garden.

"Everyone," Ann announced to the group, "this woman is writing a book on biodiversity." I was pretty sure she hadn't remembered my name. She asked Caleb and me if we wanted wine. In the spirit of journalistic virtuosity (rather than cleansing virtuosity), I said no. Caleb, wisely, said yes.

We joined the convivial guests and filled our plates. Just as I popped a giant chunk of cheese into my mouth, Ann directed her formidable gaze at me and said, "Ask away." The table got quiet and all eyes turned toward me. I tried to quickly swallow the lump of cheese as I pulled out a pen and my two pages of notes. I can't remember which of the jargon-filled questions I started with, but I know I barely managed to get through the first one. As Ann talked about the decency of Two Buck Chuck and the idea of mixing two bad wines together to create something good ("Try it, it works"), I quickly realized she didn't need erudite queries. "The Wine Aroma Wheel," she said, "was created to democratize the experience of drinking wine."

When Ann joined the UC Davis Department of Viticulture and Enology back in 1974, she explained, "wine experts were using terms like 'round' and 'balanced.' As a scientist, it drove me crazy. It created this unnecessary distance and mystique. I wanted to create a language that would help people understand and express what they were drinking without posturing. And help them understand there is no right or wrong way to drink."

This wasn't easy. We all have memories of learning how to fit wooden blocks into square and round holes as toddlers, cataloging items according to their visual similarities. But organizing by smell isn't something most of us do. It took Ann and her colleagues years to compare and systematize the wine aromas—and every sensory wheel that has come after has built on that foundational work.

These aroma and taste guides are intended to be training wheels, not touchstones. They're meant to give us support as we gradually move beyond simply knowing we prefer a wine to helping us more deeply understand and articulate why. After we get comfortable, we can—and should—set the wheels aside and journey to wherever our senses take us. But at the start, they're invaluable. "Aroma is the starting point," Ann explained. "It's the most definitive thing in wine. But the only place where we can pull these things apart is in the lab. In life, we lump taste and smell together because, when we put wine in our mouth, that's where our brain locates the experience."

This lumping together is called "taste-referred olfaction." When you put a strawberry in your mouth, for example, your brain is saying it *tastes* the fruit, but, physiologically, it *smells* it. So though we call aromas in wine "the nose," the magic happens largely in the brain. The brain changes sensation into perception: It is the place where everything we've taken in through our senses becomes what we know.

Smell, the foundational experience in wine, happens in the olfactory bulb, part of the brain's limbic system that processes memory and emotion. It also occurs in the olfactory cortex, an area that helps us identify what we smell and decide how to respond to what we're inhaling. Smell is paramount. Not just in wine—in life.

Despite 400,000 years of evolution, we *Homo sapiens* are still pretty feral, navigating the world through our noses in much the same way as wild animals. We may not actively notice, but our nose and olfactory system are one of the most important ways we distinguish between what can help or harm us.

Olfactory researcher Andreas Keller, leader of the Smell Study at Rockefeller University, thinks smell has diminished in importance as we've evolved into upright beings whose noses are farther away from the ground. Equally damaging, according to Keller, are refrigeration (which dulls odors) plus deodorants and air fresheners (which hide them). Then there's what our ancestors would consider a near obsession with personal hygiene—from daily showers to signature scents.[8] Yet, despite these interferences, smell is still our most acute sense.

"Humans can discriminate several million different colors and almost half a million different tones," Keller and his co-authors revealed in a recent study. But their conservative estimate of how many smells we can distinguish? An impressive 1 trillion.[9]

"Smell was the first of our senses," explains author Diane Ackerman, "and it was so successful that in time the small lump of olfactory tissue atop the nerve cord grew into a brain. Our cerebral hemispheres were originally buds from the olfactory stalks. We *think* because we *smelled*."[10]

And we never stop. "Cover your eyes and you will stop seeing," she says. "Cover your ears and you will stop hearing, but if you cover your nose and try to stop smelling, you will die."[11] Understanding the way smell functions helps us navigate every scent we encounter, in every breath. It's why the largest family of genes in our body is dedicated to decoding smell.[12]

Our smell, or olfactory, receptors are proteins embedded in the nerve cells in the nasal cavity inside and behind our nose. They're also found in our heart and lungs, liver and kidneys, even sperm and skin.[13]

Each nerve cell, or neuron, is covered in one type of smell receptor, which then connects with an odor molecule. The common belief used to be that smell receptors connect to a single odor the way a key fits into a lock. This, asserted Scottish researcher Robert Moncrieff back in 1949, was how we managed the constant onslaught of smells: Some odor molecules fit into the receptor; others floated on by.[14]

Theorists have now revised that argument. Odor detection isn't because of a single lock and key, they say, but the result of a combination of activated receptors working together in what's called a "binding pocket." One end of a smell receptor projects outside of the cell and connects with the odor molecules we come into contact with; the other end projects inside of the cell and carries messages to the front of our brain.

This process is astonishing. Think of a ride on a subway or a family dinner—and the hundreds of odors each of these scenarios emits. Subway smells are a full-on assault. Dinner is a bit tamer but, from an olfactory perspective, is still a bombardment. There are smells buried in the carpet, in the cleaner used to wipe off the table and in the flowers on the windowsill. There are the odors of the meowing cat and barking dog, and maybe even the scent of a leather purse or briefcase stashed on a chair, or apples and oranges set in a fruit bowl. Add to this the smell of worn clothing and warm bodies: breath, armpits, feet. How do we manage to filter all of those smells out and isolate the ones that tell us the roast in the oven—the roast that, through our triggered limbic system, takes us all the way back to Sunday dinners Mom used to make—is nearly done and it's time to crack open the bottle of wine? We owe this minor miracle to the brilliance of our olfactory system.

There are nearly 400 olfactory receptor genes from which our neurons can choose. This is what allows us to distinguish between as many different odors as possible—juice from wine, apple from pear, cat from dog.[15] When we inhale and draw in a smell, a wave of related odor receptors is activated. That's how we register familiar smells and categorize new ones.

In wine, we smell before we taste. Inhalation is something we do automatically, but it's amplified when we draw a glass toward our mouth. When we slow this experience down and make it conscious, we can start to pull apart what Ann does in the lab: distinguish taste from smell and better understand why we prefer one wine over another. As we inhale, we draw scents into the retronasal cavity in the back of our mouth and up through our nasal passages—the taste-referred olfaction Ann referenced.

This experience starts with all the events that preceded it. When we smell something for the first time, our brain builds a link between the scent and the time and place in which we encountered it. The same smell attached to a different story might change that experience: While the smell of the ocean makes a diver happy, it might make someone who gets seasick a little nauseated.

We taste by association, Ann explained as she refilled Caleb's glass with wine and mine with water. "We say a wine *tastes* like cherries or strawberries because we can only communicate what we already know." And we can only communicate with the language we have. English isn't a smell-based language, but psychologist Asifa Majid and linguist Niclas Burenhult have done extensive research into the links between language and smell and found that isn't always the case. The Jahai tribe from the Belum forest of Malaysia speaks in a language rich with smell descriptors. In a comparative study between American English language speakers and Jahai language speakers, they found English speakers were "vague and inconsistent" when naming smells but spot-on when naming colors. The Jahai, however, excelled at both.[16]

When we take our first sip of wine, smells mix with tastes. There are five tastes: sweet, salty, bitter, sour and umami, plus an innate receptivity to fat (which also has its own taste). Sweet, bitter and sour dominate in wine, though Caleb confirms umami and saltiness may come through. Our mouths also register the feel of the wine: Reds that are high in tannins (polyphenols that are also found in persimmons and the skins of nuts) can make our mouth feel astringent or dry. Chardonnays that undergo malolactic fermentation (the conversion of tart, bright malic acid into softer lactic acid) come across as creamy or buttery. It's incredible: In an instant, these experiences process through our brain's orbitofrontal cortex to become flavor.

The good news, Caleb explained, is that "usually the nose [smell] of the wine matches the taste. They line up." What was even better was Ann's parting comment as she walked Caleb and me to the door:

"People often ask me what makes a good wine taster. They think I'm going to tell them it has something to do with sensory acumen, but I tell them it's someone who takes the time to savor—a person who can focus."

When I fell in love with that glass of Trousseau Gris, I wasn't thinking about Ann's aroma wheel. I wasn't aware of my retronasal cavity, let alone my olfactory receptors. The only thing I was conscious of during that first transcendent sip was the Trousseau Gris—a remarkable grape that almost never made it to my lips.

Trousseau Gris is a white wine grape native to Jura, a small wine region in eastern France. Researchers speculate that Cistercian monks transplanted the variety from its place of origin to southwest France in the 12th century. Then, after five centuries of production, Trousseau Gris disappeared for many of the same reasons crops disappear today, including temperature (a harsh winter in 1709 wiped out a number of vineyards), disease (specifically, sensitivity to the fungus *Botrytis cinerea*) and changing tastes and production practices (Trousseau Gris didn't work well with the Charentais wine distillation process that was in vogue at the time).[17]

Growers did then what growers have always done: They replanted areas that had lost vines with varieties that were heartier and better suited to popular tastes and methods of production. As a result, the Trousseau Gris grape nearly went extinct.

Trousseau Gris was planted throughout California in the early 20th century. However, by the 1980s, it began to disappear. Now, in North America, the only commercially viable plot of Trousseau Gris is located on 10 acres in the Russian River Valley, a cool, rain-fed area known for its Pinot Noir and Chardonnay. The steward of that land, and those grapes, is Peter Fanucchi.

Caleb and I visited Peter at Fanucchi Vineyards on a surprisingly warm morning in early February. Dressed in a T-shirt featuring the American flag, Peter greeted us with a warm smile and a hearty handshake. Within minutes, we were all sitting before bare grapevines, holding in our hands glasses of the vine's last harvest.

Peter is conservative in every sense of the word. He is the polar opposite of the California hippie vineyard owner I had envisioned, and the polar opposite of me—on the First Amendment, the Second Amendment, women's reproductive rights, our president, everything. Well, almost everything. There we were, six months after my first taste of Trousseau Gris, talking about the wine we both loved. The wine that wouldn't exist were it not for him.

As we sipped the wine, Peter explained that the Fanucchis' cultivation of Trousseau Gris was mainly due to luck. "Back in 1981," he said, "my dad asked me which of three grapes to grow. The choices were Gewürztraminer, Chardonnay or what, at the time, we were told was Grey Riesling. It was actually Trousseau Gris. My dad was a little afraid of Gewürztraminer because he thought people would have a hard time pronouncing it, so we ruled that one out. I was a kid and didn't drink wine, but I loved the sweetness of what we thought were Grey Riesling grapes, so that's what I chose."

The vines were bred from cuttings from Wente Vineyards—the oldest continuously operating, family-owned winery in California—and from a grape collection managed by the U.S. Department of Agriculture and the University of California, Davis.

Back in the 1980s, when the California wine industry was establishing its identity and reputation, any white grape variety that wasn't a Chardonnay or a Sauvignon Blanc was pulled out or used to bulk up larger blends. "This extended through the early 1990s," Peter said. "The only white variety making any money was Chardonnay. But I loved Trousseau Gris. I knew I had to make money by selling the grape or the wine. I had to show the grape was worth something in order to get the price—and justify keeping it in the ground."

Even under the best circumstances, these decisions are hard. It's not easy to stick with a less popular (and, therefore, less lucrative) grape—or to pull out the source of your income and wait three years for your new investment to bear fruit.

This challenge also extended to Peter's Zinfandel: "I kept it when it wasn't fashionable or lucrative by managing to get folks into long-term contracts for various blends. Gallo blended it into White Zinfandel; Piper Sonoma bought up all the fruit during another period and put it into their sparkling wine. There were times when we certainly could have torn the vines out, but we didn't. We kept them," he said. "It was partly financial because it was never really feasible for us to be out of production—and it was partly because I believed in them."

Peter devoted his time, energy and agricultural inputs (such as water and fertilizer) into keeping his vines alive. He wouldn't let them disappear: "I know the seasons. I know this land. I know this grape. I could be one of many growing Chardonnays—or one of the few growing Trousseau Gris."

This practice is what's known as in situ—Latin for "in place"—conservation. The two main methods are to save and protect what's already growing in the wild ("in situ conservation in the wild") and to sustain a plant or an animal by actively growing or raising it ("in situ conservation on-farm"). These efforts are critical because they are dynamic, constantly responding to changes in the environment and in our culture.

But it's only possible if farmers are able to commit to it. In order to fully appreciate the intoxicating biodiversity Peter conserved, we need to rewind back to grade school biology and taxonomy, the organization of all living things—generally categorized as kingdom, phylum, class, order, family, genus, species, variety.[18] These classifications were developed by Swedish naturalist Carolus Linnaeus, considered the father of taxonomy, who, in the 1700s, developed the classification system used for naming all organisms. Taxonomy clusters plants and animals together according to qualities we can observe (such as shape and size), as well as internal structural and genetic traits. It also includes a categorization often connected to geographical location: variety.

The Trousseau Gris grape variety is part of the *Vitis vinifera* species, in the same Vitaceae family of flowering vines that includes the Virginia creeper. Wine can also come from *Vitis labrusca*, *Vitis rotundifolia* and other species, but 90 percent of wine grapes are *Vitis vinifera*.

Although California is one of the largest wine producers in the world, the majority of its vineyards are dedicated to growing just eight grape varieties: Cabernet Sauvignon, Chardonnay, Merlot, Pinot Noir, Sauvignon Blanc, Syrah, Zinfandel and Pinot Gris. These same vineyards are also dedicated to growing varieties many of us have never heard of, namely French Colombard, which is used in cheap sparkling wine and to make bulk wine more acidic, and Rubired, which is used mostly for coloring and grape concentrate.[19] While the Wine Institute of California's website proclaims the state is "a perfect place to grow nearly every kind of grape,"[20] as Caleb likes to remind me, what is actually grown "is a lot of sameness."

If we can grow nearly every grape in California, then why don't we? That requires a history lesson.

The first viable grapevines were planted in California in 1779 by a group of Spanish clergy led by Father Junipero Serra. They were likely brought from Spain to Mexico in the mid-1500s, and then on to Texas and New Mexico in the 1600s—and were the first *Vitis vinifera* species grown in the United States. The genetics are unclear, but what we do know is that, as the vines made their journey from Europe through North America, they responded to their environments and the vines around them and soon became the thick-trunked, hearty variety used for Communion and as table wine for the Church.[21] (Jesus drank wine with meals, too.[22]) Because they were established by Catholic missionaries, these fruits of the vine came to be known as Mission grapes and, until about 1850, represented the totality of wine grapes grown in California.

Although there were vineyards in the southern part of the state, in Los Angeles and Anaheim, the most popular wines at the time came from the east and Midwest, not California. But the gold rush of the mid-1800s changed that. As the state's population and coffers swelled, vine plantings more than doubled and helped bolster vineyards and wineries in and around Napa Valley and Sonoma County.[23] The gold rush begat a wine *gush*: Winemakers in the United States (the New World) and Europe (the Old World) traded vine cuttings as production continued to expand.

Then, in 1863, disaster struck in the form of a tiny, aphid-like insect called phylloxera.[24] These greenish-yellow bugs feed on the roots and leaves of grapevines and eventually kill the plant. The first recorded case of phylloxera infestation was in Provence, France. It quickly spread through the wine capital of the world during what became known as the Great French Wine Blight, eventually reaching Italy, Germany and other wine-producing regions.[25]

At the time, people didn't understand what was happening. The Italians thought it was the wrath of God because of humanity's sins (which presumably included drinking copious amounts of wine). France was so desperate that the minister of agriculture offered a monetary prize of 300,000 gold francs for a solution. People went berserk, suggesting remedies that included "cows' urine, powdered tobacco, walnut leaves, crushed bones dissolved in sulphuric acid, a cocktail of whale oil and petrol, hot sealing wax applied to pruning lesions, potassium sulphide dissolved in human urine, volcanic ash from Pompeii, marching bands, [and] douches of elder-leaf tea."[26]

The mystery wasn't solved until 1868, after nearly 40 percent of French vineyards were already destroyed. French botanist Jules Émile Planchon and his colleagues did a comparative analysis of healthy, dying and dead vines and found the pest phylloxera sucking on the roots of the live plants. "It was discovered that by the time the vines were observed dying, the aphids had already moved on to their next meal."[27]

The reason this infestation happened so quickly—and resulted in such devastation—is because phylloxera, an insect indigenous to North America, likely hitched a ride to Europe on the boots of a vintner or in the roots of a plant. Because European varieties had never come in contact with phylloxera, they had no resistance to it. American varieties, however, did. The roots of American grapevines secrete a sap that chokes young (nymph) phylloxera as they try to suck out the sap. This is why Old World varieties that had been sent from Europe and planted on American soil also suffered.

"The solution lay in the cause," says viticulturist Christopher Bland. "American root-stock, which had harboured the disease, was also resistant to it."[28] By the 1890s, all French vineyards were replanted with French scions (grape shoots with buds) that were inserted, or grafted, into phylloxera-resistant American rootstock. This effort was replicated throughout Europe and on all Old World vines in the United States. America nearly destroyed—but ultimately saved—the world's wine.

This risk, what scientists call "pathogen pressure," hasn't gone away. Those sap-sucking menaces still exist, and there's no way to get rid of them other than making vines less welcoming. Pesticides don't work because they can't really penetrate the heavy soils that phylloxera love;[29] resistant rootstocks are currently our only solution.

Not long after recuperating from the phylloxera plague—just as it was finally achieving global recognition—the American wine industry (led by California) encountered another setback: Prohibition, the constitutional ban on the production, transportation and sale of alcohol. Some vineyards survived by converting to table grape or grape juice production, and a few others managed by producing wine for the Church, but most of the industry went under. By the time Prohibition ended in 1933, only 140 wineries were left standing.[30]

It wasn't until the 1960s that the California wine industry finally reestablished itself. Big wineries started to develop, and the signature tastes that have come to define the region emerged. This transformation was, in large part, due to winemaker Robert Mondavi. Mondavi focused on selling wines as varietals—wines made from single varieties of grapes. This effort helped build the reputation of select grapes like Chardonnay and Cabernet Sauvignon over blends. Mondavi also committed to adopting techniques he saw in France, such as aging wine in French oak barrels rather than California redwood, which, Caleb explains, had "a tremendous effect on flavor."

Over the next decade, Cabernet Sauvignon, Chardonnay and other French varieties started to thrive in Napa Valley and Sonoma County, just as they had in Bordeaux and Burgundy. They advanced to such a degree that, in a blind tasting in Paris (famously known as the Judgment of Paris), American reds and whites—including a Cabernet Sauvignon vintage produced from vines that were just three years old[31]—bested their French counterparts.[32] In a tasting rematch 30 years later, American reds, once again, won "by a nose."[33]

These accolades were largely a blessing. They grew California into the largest wine-producing state in the U.S. and the fourth largest producer in the world. And they helped define what the world considers

American wine to be: big, bold reds and creamy, vanilla whites. But this definition has also limited what California wine is—and can be. Thirty companies are responsible for about 90 percent of the wine sold in America. Three companies—Gallo, Diageo and Constellation Brands, which now owns Robert Mondavi Winery—represent over half of domestic sales.[34]

"Until pretty recently, most winegrowers and winemakers have been scared to take a risk," Caleb explained as we walked through Peter's vineyard. "People want something that tastes like what they know the Russian River Valley or Napa to be."

That's what industrialization does. It builds the brand. It provides greater consistency in quality and availability and, possibly, a lower price. But grapes are living organisms. In order to ensure we get the same taste out of every bottle, something has to be manipulated, compromised or added.

Because winemakers aren't legally required to disclose all ingredients, and wine is, essentially, fermented grape juice, we tend to think of wine as just one thing: grapes. However, there are a host of inputs that can be added: Animal bones, fish bladders, milk proteins and other protein-based additives are used to filter and clarify; silicone oil reduces frothing during fermentation; and sulfur stabilizes and sterilizes.[35]

It doesn't have to be this way. Grapes are self-contained units of winemaking magic. They hold everything needed to make wine—yeast, sugar and water—right there in the grape. Yeast occurs naturally on the skins of grapes, while the sugar forms as the grapes ripen. Wine is produced once the yeast feeds on these sugars.

But in order to achieve the scale and consistency consumers demand, industrial winemakers have to include additional ingredients, while their suppliers have to irrigate and fortify the soil with nutrients to keep growing the same grapes over and over. As a result, the water tables drop and the land never gets to rest. Peter explained, as we drove from one part of his vineyard to another, "Industrial wine has reshaped what the Russian River Valley grows." Caleb agreed, chiming in: "There isn't a lot of room for experimentation, variation in vintage or exploration of tastes."

This is why the wine landscape is dominated by so few grapes; 66 percent of the world's wines come from just 35 wine grapes.[36] It is also why anything grown outside of those dominant grapes is so special. Peter grows what are known as "underutilized" grapes, including Trousseau Gris. "Underutilized" refers to crops that could be grown but, for a variety of reasons (be it agronomic, genetic, economic or cultural), no longer are.[37]

This, of course, isn't limited to grapes. Fifty-nine percent of America's farmland is dedicated to growing just a handful of commodity crops. Corn alone (what's used for grain, not animal feed) accounts for about one-fourth of land under cultivation.[38] The implications of this are huge, not only because it edges us closer to the monotonous global standard diet, but because it affects winegrowers, winemakers, farmers and producers all over the world whose diverse plantings are systematically replaced by homogenous varieties—and whose ingenuity is constrained by the rigid adherence to a small group of popular crops.

It makes life a lot less interesting and delicious for all of us.

Another reason Trousseau Gris never made it into the spotlight was because of its questionable reputation, which, in grape terms, translates into being misidentified as a neutral grape. "If you look at the old texts," Peter explained, "it says this grape is a neutral fruit with no aromatics. [Wine journalist and critic] Jancis Robinson called it a 'neutral grape with no character.' But that's because people picked it too early."

When Peter first made wine from these "neutral" grapes, he discovered the opposite of what had been written. Trousseau Gris had "great aromatics and great fruit, and—with gentle pressing—great flavor. The grape doesn't like to let its juice come out, so people would press hard on the thick skins. But when it's pressed too hard, it gets acidic. That made the wine bitter, so then people thought the grape was bitter." (This is also what Scott Schultz, the winemaker behind Jolie-Laide Trousseau Gris, discovered when he used Peter's grapes.)

Peter's determination reminds me anything is possible. He saved this crop when there was no established market for it. He nurtured it because he loved it. While I do not share his political beliefs, I'm in awe of his commitment to biodiversity and efforts to "grow in harmony with nature." To Peter, this means avoiding synthetic pesticides or fertilizers: "The flavor, the distinctiveness, is about an environment in which the fruit grows—and it's reflected in the fruit. I work this land and I live here." He then added with a smile, "And I want my tombstone to read 'It's not a neutral grape.'"

As Caleb, Peter and I wound our way through the 20-acre vineyard, Peter talked a lot about intimacy. "Big scale winemakers can't learn their locations. Nearly all of this work," he said, gesturing broadly, "was done by me." I studied his thick, clean hands; his middle-aged belly; his dazzling blue eyes and salt-and-pepper beard. "This vintage is about being patient, something I have done my whole life."

Big Wine can't give us this intimacy, these relationships. This is why we need underutilized varieties to flourish.

After showing us the vines closest to his house, Peter then drove us to a different part of the vineyard. Because it was winter, the terrain looked much the same, but I could feel the shifts in temperature and humidity—the microclimates that are one of the defining characteristics of the Russian River Valley. "The best conditions for wine are stress," Peter explained as we climbed out of his truck. "Rocky soil, just enough water ... what California has." (Although California's ongoing drought will likely change this.) I have returned to that comment many times and mapped it onto all parts of my life: The best wine comes from stressed vines.

When Caleb and I visited Peter, his vines were dormant and grapeless, but I was adamant about seeing them. I wanted to stand on the ground where the grapes were grown. Peter and Caleb understood. They knew that, though nothing was in bloom above ground, there was a lot going on below in the soil, the top layer of the earth's surface where biodiversity begins.

To be clear, soil isn't dirt: "Soil is what plants grow in. Dirt is what you get beneath your finger nails."[39] Dirt might support plants, but it doesn't have to, because it's essentially anything that comes from the ground and has a color and texture similar to soil. Potting mixes are a great example. They are soilless—made up of moss, bark and minerals—

and usually sterilized. This kills any microorganisms that might carry disease, but also kills ones that help plants grow.

Soil is something else; it is something precious, home to one-fourth of the world's biodiversity and full of life. This mixture of minerals, water, air, organic matter (decayed plants and animals) and microorganisms is what sustains life on earth—and is the place from where 95 percent of our food originates.[40] It's as fundamental to our existence as water or air.

And soils, like people, are transformed by the company they keep. They develop slowly, over time. Top soil, the upper, most productive layer of soil, takes over 500 years to form. That precious inch—and all remaining layers of soil—evolves in response to the climate, terrain and organisms with which it comes in contact. So while we may think of soils as static and constant, they're not. Soils are living entities: They are born, they breathe, they age and they can die. They *are* dying.

The Food and Agriculture Organization of the United Nations (FAO) estimates over one-third of all soils are degraded, meaning they are eroded, compacted, polluted or less fertile.[41] This is due to a changing climate and the relentless pressure we put on our land; heavy-tillage agriculture with limited, if any, crop rotation is one of our biggest problems. According to the 2015 *Soil Atlas*, "We are using the world's soils as if they were inexhaustible, continually withdrawing from an account, but never paying in."[42] We are wearing the soil out.

It's impossible to grow good food in bad soil. Unless we extend our cares about food all the way down to roots in the soil, our global, per person amount of productive agricultural land, by 2050, will be one-fourth of what it was in 1960.[43]

"The Earth's 'skin' is not one soil, but many soils—each with its own story," Smithsonian experts tell us. "Tens of thousands of different soils cover the continents."[44] In fact, scientists have identified over 70,000 kinds of soil in the United States alone.[45] They are bound to place, an inextricable part of what is called terroir (pronounced "ter-wah").

"In a very basic sense, terroir is the expression of a physical place through the flavors and quality of a wine [and other drinks and foods] ... including the climate of the vineyard where the grapes were grown, the soil, the geography, as well as the characteristics of the grape variety itself," writes wine blogger Becca Yeamans-Irwin.[46] Terroir, in other words, is the taste of place.

The geography of the Russian River Valley has been shaped by volcanic eruptions, shifting tectonic plates, plus the water that flows down from the Sonoma Mountains that has resulted in an area characterized by three distinct soils. Peter's Trousseau Gris grapes grow in a rich soil made of clay and sand—a sandstone loam known as Goldridge soil. His vineyard sits in an area thick with morning fog that tends to burn off by midday. The cool morning and night temperatures mean his grapes ripen slowly, which allows the sugars in the fruit to develop fully.

I breathed in the cool air as I reached down and touched the loose, dark soil and barren vines. This was the terroir of the wine I loved.

"An ideal wine will be able to share flavors that tell of the land, the weather that year, all the decisions about the vines," Caleb explained as we returned to his car. But the information has to be received. "Pay attention to the wine," he urged. "Listen to it. It will open you up to a broader spectrum of tastes and styles. Pay enough attention to the grapes to let them speak for themselves."

In vino veritas: In wine, there is truth.

II. ON THE VINE

For centuries, wine has symbolized something sacred: blood, truth and transformation. "What batters you becomes your strength," wrote poet Rainer Maria Rilke in his work *Sonnet XXIX*. "What is it like, such intensity of pain? / If the drink is bitter, turn yourself to wine."[47]

There is something magical in recognizing the alchemy not only of foods (grain to bread, water to wine) but of ourselves. It is so easy to feel despair for our food supply and for the world. The grace, as I learned from Snow Barlow, is in finding hope: in being like the grape and turning ourselves to wine.

Snow Barlow is a 70-year-old Renaissance man: a former professor of horticulture and viticulture at the University of Melbourne in Australia, the current chairman of the Victorian Endowment for Science, Knowledge and Innovation *and* a winemaker. When we first met at a formal event for academics at the university (where I had been invited to be a visiting scholar), I introduced myself as Simran. He smiled and immediately asked what my friends called me. "Some call me Simi," I said, taken aback by the question. From that moment on, that's what he's called me. Instant friendship.

In our first interview—over handmade Italian pasta on Lygon Street in the heart of Melbourne's Little Italy—he held up a glass of wine and proclaimed, "Let's be like a grape." I was game.

When Snow talks, he leans in close, almost as if he's sharing a secret. This makes even the most mind-numbing data seem intimate and interesting. "If it's too cold," he began, "you never quite make it to maturity. If it's too hot, you mature too fast. There's a lot of sugar in grapes, and the delicate flavor compounds need time to accumulate. When the plant is grown in the right place—the climate is right, the soil is right—it expresses the characteristics that make it unique. Taste is manufactured in the plant, not the soil, Simi, but the soil is the starting point. It influences the plant to produce certain compounds that are delicious."

Snow said this to me during the summer of 2014—at that point, the hottest year in recorded history. Much of his research has focused on how climate change impacts the growth of grapes and the taste of wine. "The reason we go to great lengths to find the exact place to grow grapes," he explained, "is because only certain places express the full genetic potential of exquisite flavors." But now those places are changing, and we don't know exactly where the best flavors will end up. Climate change is changing terroir.

Some are placing bets on a flourishing industry in places like England and China.[48] But, as Caleb explained in a later email, with the introduction of new clones and genetically engineered rootstocks and yeasts, anything seems possible—including wines from Florida and Thailand.

Industrial wines are already manipulated and tweaked: "For wines outside of the best quality," he wrote, "there will continue to be options to fix things in the cellar, like adding acid to overripe grapes. The ordinary quality level will remain, as will the sort of bulk wine level for supermarkets." But the good stuff will change.

"The highest quality wine production will push north," Caleb continued. "Some of the best wines are made on the edge of viability, like in the Sonoma Coast, Champagne and Burgundy. It's probably not too hard to imagine Burgundy, say, becoming an ideal location for Cabernet Sauvignon. The sort of idealized marriage between grape and land may never be the same again. It may take a long time to get grapevines adjusted and acclimatized to their new homes, the way that vines in Europe have adapted [over centuries] to their native and adoptive locales."

When I shared Caleb's insights with Snow, he agreed. "We're still being like a grape," he said with a wink as we tucked into our second pasta lunch. "Water is the currency of the grape—and of climate change. The life cycle of the grape is impacted by the availability of water and by temperature. Generally speaking, this means climate change impacts the *quality* rather than the *productivity* of grapes. It makes some of the wines we love taste different because grape ripening is forced into hotter windows."

I took a long sip of Pinot Gris and tried to comprehend what that meant. Snow leaned in: "Hotter climates are usually deleterious to fruit composition, Simi. When the climate is too hot, the grapes can't achieve their full genetic expression." The grapes can't achieve their full grapey glory—for size, yield, duration of season or for taste. "Climate change will change the way people drink," Snow continued. "Our wines won't just move north; they will lose flavor. The old, stressed vines that make the best wine will transform—or disappear."

These changes are already being felt in wine regions around the world. The droughts of 2013 and 2014, for example, have been the driest seasons ever documented in some parts of California. It isn't clear exactly how this will impact grape harvests,[49] but, thus far, the agricultural sector as a whole (California grows nearly half of America's domestic fruits, nuts and vegetables[50]) has lost $1.5 billion due to drought and other weather-related challenges.[51]

Globally, a warming planet will bring more opportunities to grow food in colder places; however, the latest Intergovernmental Panel on Climate Change (IPCC) report reaffirms that the world as a whole will experience more food loss than gain.[52]

Although we don't know exactly what will happen, we do know agriculture is changing and will continue to change. It will impact the biodiversity in wine for worse and, in some instances, for better. This is why we need to grow a diversity of grapes and other crops. Diversity enables us to respond to whatever circumstances arise; it makes us more resilient.

"The genetic base for grapes is quite narrow," Snow said during our final lunch. "It's a bit like a pyramid. And there are good, industrial reasons for that pyramid being pointy, like high yield and stress resistance. We simply can't improve hundreds of varieties."

Indeed, but the challenge is that we don't know which varieties that aren't being grown widely will be important further down the line. We don't know which ones will have greater drought tolerance or resistance to pests or disease. Think back to when the world was trying to grapple with phylloxera. Most times, we don't know what to guard against until it happens.

As we see with California's ongoing drought, there's still a lot to learn. "We don't understand which genes allow a plant to read the environmental cues of seasons, temperature and day length," Snow explained. "We don't understand, for the most part, how these genes stimulate a plant to flower or ripen, and what chemical signals they use to do this." He paused and threw up his hands. "And then at the end of the season, how do they sense changes in day length and temperature to know when to go dormant to avoid being injured by the first freeze? These are all things we have to understand to build the capacity for resistance to climate change."

And we don't know what we'll grow to love, such as the non-neutral grape Trousseau Gris. If those grapes, or any other crops, aren't cultivated on farms or accessible and identifiable in the wild, then we lose them. That is, unless they're saved in backup collections stored *ex situ* ("out of place").

The most popular types of *ex situ* collections are seed banks, but we conserve all kinds of genetic material, collectively known as germplasm. This includes honeybee sperm, fish eggs, cattle embryos, microbes and tissue needed to propagate plants. Crops that aren't grown from seeds are often stored as living plants in open-field genebanks—"germplasm repositories." Animals are preserved *ex situ* on teaching farms and in some zoos.

Nearly half of all material (known as "accessions") in the world's genebanks are cereals. Legumes, such as beans and peas, make up about 15 percent of accessions, followed by fruits and forage crops eaten by livestock, which account for roughly 10 percent of stored material.[53]

Germplasm repositories are typically located in climates that have the greatest potential to grow the widest range of whatever is being conserved, so, not surprisingly, the largest *ex situ* collection of grapes in

North America is in California. Officially known as the USDA National Clonal Germplasm Repository, the genebank in Davis is devoted to tree fruits, nut crops and grapes. It is also the place that supplied a portion of the Trousseau Gris that Peter Fanucchi's father planted back in 1981. Horticulturalist Bernie Prins is its manager.

Bernie is super-brainy, super-accessible and super-cute in that brainy, accessible sort of way, with tousled hair, thick glasses and an endearing gap between his two front teeth. "I'm a very weird vineyard grower," he announced soon after we met. "Pay less attention to the fruit and focus on the vines. My job is to keep all of them alive."

By "all of them" Bernie means 6,000 vines of over 3,600 accessions of grapes, including more than 1,200 types of European wine, table and wild grapes. "It's the most extensive collection of wild wine species in the world," he explained. "And the single biggest collection of Greek varieties anywhere, including Greece."

Bernie means it when he says grapes aren't his focus. The vines at the genebank are dotted with unharvested, shriveled-up grapes. When Bernie, Caleb and Peter (who joined us on our tour of the genebank) weren't looking, I pulled a few Trousseau Gris *raisins* off the vine. They tasted moldy.

The goal of *ex situ* collections is to maintain as much genetic diversity as possible. "Half the grapes don't taste good. But that's expected. We don't keep them only for taste but also for things like resistance to disease or cold." Bernie and his colleagues also grow the wild descendants, known as "crop wild relatives," of not only hundreds of grapes but also about 3,500 genetically different fruit and nut trees, including peach, plum and apricot, most of which are grown in California. (The state produces more than 90 percent of tree nuts and about half of all fruit harvested in America.[54])

Crop wild relatives are the plants from which a crop was domesticated, or a closely related species. Wild relatives are essential to maintain because one in five plant species worldwide and one in three native plant species in the United States is endangered by climate change, pests, disease and loss of habitats.[55] By sustaining a wide range of diversity in dynamic collections that respond to environmental challenges, we give breeders and growers more opportunities and alternatives for keeping the wine industry (and agriculture as a whole) afloat.

Growers in Sonoma County and Napa Valley, for example, are already starting to switch to grapes that are more drought tolerant and adaptive to warming temperatures. But there's a limited amount of genetic variation in what we currently grow, which is usually only what we can sell. This is why *ex situ* collections are crucial. The thousands of grapes that are here include noncommercial and foreign varieties, plus crop wild relatives; breeders can tap into a larger pool of genetic traits beyond what might be commercially available.

Some wild relatives are unrecognizable in terms of appearance and functionality. Oranges, for example, look and taste nothing like their wild ancestor, the key lime. Bitter apple, the wild ancestor of watermelon, has white, bitter pulp. Teosinte grass, a wild plant that originated in southern Mexico, which bears no physical resemblance to modern-day corn, is, genetically, nearly identical. It carries genes for resistance to certain diseases that damage corn, which researchers have used to develop the virus-resistant corn varieties we grow today.[56]

Preservation of crop wild relatives is critical because, though they look different, most domesticated fruits and vegetables are genetically similar (think of two sisters). *All* modern domestic tomatoes, for example, "possess no more than 5 percent of the total genetic variation present within the wild species and [indigenous] varieties."[57] Crop wild relatives, however, possess far more genetic diversity (think of your out-there distant cousins). These wild ancestors are an essential part of the family of living things.

Continuing on our tour of the genebank, Bernie pointed down a row of vines. "We can cross some of these wild grapes with grapes native to a place to increase disease resistance and give people a taste they are familiar with." Familiarity is a big deal because with investment comes expectation: We want a Pinot Noir to taste like what we think a Pinot Noir should be. Caleb chimed in, "We crave sameness, but what we also have to remember is that uniformity comes at a price. Something has to be tweaked to make a living, dynamic thing look, feel, taste and smell the same season after season, year after year. We have grown to expect this in food, but would never expect this uniformity in our garden—or in someone we love. We respond to our climate and our environment; we change with the times."

Yet, according to Bernie, "there are almost no new wine grape varieties being produced, which is in contrast to both table grapes and also almost every other fruit crop where consumers readily accept new and improved varieties." He explained, "Everyone wants to drink a 300-year-old French variety with a 300-year-old history, but it's *impossible* for a modern wine grape breeder to create a brand-new 300-year-old variety."

Why haven't we reached that point of acceptance? Why aren't we willing to expand our palates? The repository is also the biggest in the world for pistachios—hundreds of varieties are grown here—but just one dominates the market: the Kerman. We walked amid thousands of varieties of grapes, but only a fraction of them are ever made into wine. Why don't we grow a greater variety *now* as opposed for waiting for a new pest or disease? Didn't we learn from the Great French Wine Blight?

Some crops, Bernie responded, do have active breeding programs. But grapes aren't among them. "Then why don't more farmers just grow them on their own?" I asked. "They're the ones who will suffer most if their monocrop is wiped out by a pest or disease. Isn't a long-term investment better than short-term gain?"

Think of that investment, Bernie said. "That short-term run can last up to 100 years. It's much cheaper to just keep growing Kermans—or a handful of wine grapes—since that's what sells."

Peter nodded in agreement. "It's a balancing act," he said. "Struggling to try to have the income to survive the day has to be a priority or you can't invest in the future. I try to put in the time to do both, but oftentimes buying the new equipment and the new materials is put on the back burner."

This is another moment that humbles me, one of many on this journey. The future of wine, fruits and nuts is in places like this. Yet most *ex situ* collections are compromised by too little time, too little money—and too few people to do the work.

This is, in part, what prompted the creation of a global backup repository for seeds: the Svalbard Global Seed Vault, often referred to as the "Doomsday Seed Vault." As former executive director, agriculturalist Cary Fowler explained to environmental correspondent Suzanne Goldenberg, "Why did we build it? It wasn't because some apocalypse was coming. It was because we knew gene banks were losing samples, and were losing them for stupid reasons—[budget] cuts, equipment failure and human error. Prior to the seed vault we were losing diversity. I am convinced that we were losing at least a variety a day, silently. It was this kind of drip, drip, drip of extinction. We put an end to that—at least for 865,000 varieties."[58]

Whole-plant collections like Bernie's have to be continuously grown to preserve genetic biodiversity for future use. A world made sweeter with wine, peaches and almonds means supporting the work of people like Bernie and Peter, along with many others.

And this is where we, the eaters and drinkers, come in. The final form of conservation is one that touches us all. It's what I call *in vivo*, "in living," conservation: saving foods and drinks by consuming them. This term is, admittedly, a bastardization of the scientific use of the term *in vivo*, but it works—and it's necessary. Conservationists don't have a word for what we, the people, can do. They haven't considered us as part of the struggle. But we are. We all have a vested interest in what sustains us. The only way Peter can keep his 10-acre plot of Trousseau Gris going, the only way we can all ensure a world *with* wine, is to diversify—an impossible task without support from the drinking masses.

Early in our interactions, Caleb sent me this note:

> Just found this quote from Paul Draper of Ridge winery excerpted from a 1994 seminar at the Académie Internationale du Vin: "Wine is different from other alcoholic drinks, in that you have everything necessary to make it present in the raw material. It's due to this simple fact that wine, since the dawn of Western civilization, has been the essential symbol of transformation, both physical and spiritual. I sometimes think that my colleagues in the New World forget that the power and mean-

ing of wine comes from nature and from a natural process, not from man or an industrial process."[59]

This idea of transformation has been an important part of these last three years. The foods I selected went through many permutations and were chosen with great care after countless hours of consideration (or, some would say, obsession). They're all foods that stimulate, soothe and nourish us in multiple ways. My sister reminded me they're all considered vices and told me to keep reaching for pleasure; my friend Stephen pointed out they are all fermented. The pleasure I was aware of, but I didn't anticipate the connection to fermentation. The only foods that sprang to mind when I thought of fermentation were sauerkraut, pickles, yogurt and kimchee. Not anymore. Bread, wine, coffee, chocolate and beer—they're all fermented. As Sandor Katz details in *The Art of Fermentation: An In-Depth Exploration of Essential Concepts and Processes from Around the World*, starters used to trigger fermentation are called "cultures." "Culture comes from the Latin *cultura*, a form of *colere*, 'to cultivate.' Our cultivation of the land and its creatures—plants, animals, fungi and bacteria—is essential to culture. Reclaiming our food and our participation in cultivation is a means of cultural revival."[60]

For millennia, people have dedicated their lives to perfecting the beverage nature has brought forth. Pick, crush, ferment, age, bottle: We taste the decisions in the glass. Winemaker Scott Schultz of Jolie-Laide Wines is a member of this lineage; one who helped trigger my transformation.

"In 2010, I made Trousseau Gris for the first time," Scott told me on the phone after my wine epiphany at Camino. "I was looking for an inexpensive white, something that had bright acidity and aroma." That led him to Peter Fanucchi's vineyard. "When Trousseau Gris grows, there's a lot of pigment in the skin. I wanted that density and texture, that intrigue, so I decided to leave the skins on for a few days as a cold soak, which made it this pinky orange color." It worked: His wine looked like a sunset.

Scott talks about wine the way chefs talk about food—as a full sensory experience. He came to California to work in Napa Valley at

Bouchon, one of the country's most delicious French bistros. "I started meeting all these winemakers and volunteered for harvest back in 2007, on my days off from the restaurant. I loved it. It was a lot like the craziness in the kitchen, but we were making wine, which I thought was so cool." That exposure is what inspired Scott to commit to making wine full-time.

"When I first made Trousseau Gris," he admitted, "I wasn't sure if I liked it because *I* made it or because it was good."

Scott dropped off bottles of that experimental Trousseau Gris to a few restaurants he respected, including a bottle to Caleb at Camino. "I remember Scott was really apologetic that he was dropping it off in the middle of service," Caleb said. "And then I tasted it. It was creamy and had these great peach nectar and herbal notes. We use a lot of herbs at the restaurant, so it being herbaceous wasn't a problem, but it was a really different kind of California wine. I thought to myself, 'This is really weird—and it's fucking great.' It was a wine unabashedly itself."

Jolie laide is a French expression that means unconventionally attractive, and it describes perfectly the essence of this wine and other wines Scott makes. "Greenness is often considered a flaw in wine," Caleb explained. "It means your grape isn't ripe. But here, in this wine, it's something else. It's not a flaw—it's appealing. It brings a complexity and earthiness."

This "greenness" Caleb described is in Scott's very makeup. He's down-to-earth, real. And he makes the elusive qualities of a wine somehow tangible with beautiful, unconventional wine labels that change every year "because the wine is different every year." He recognizes it's a living entity: Before it's made, it's grown.

Chemically speaking, the difference between bottles of wine is small. But something in the Jolie-Laide Trousseau Gris captivated Caleb and, subsequently, captivated me. When I asked Scott to describe the wine in person, he was vague. A week later, I asked him via text message, explaining I was celebrating the birth of my second nephew with the Trousseau Gris he had gifted me. I was hoping a baby would inspire some feedback.

"Be hypnotized by the juice," he wrote. "Everything is fleeting, so get it while you can."

"Please tell me what you love about the Trousseau Gris," I responded.

"It's never about me," he wrote. "I hope each owns his or her experience."

I wasn't getting anywhere. I asked what he was drinking and asked him to describe it. He did so beautifully.

"Okay," I typed, "now pretend the Trousseau Gris wasn't yours. You told me you make wine you want to drink. How would you describe it?"

"Not fair," he responded.

"Just humor me," I pecked out. "I find this naked lady on the wine label. She's bold, strong and unapologetic. Revealed, fully. But there is an unmistakable sweetness to her. I want to be like her. Maybe 'sweetness' is the wrong word. She's floral, but also sharp. She contains multitudes."

"As far as the girl," he wrote, "for me it was supposed to resonate as simple, naked, untouched. Which says the same thing about the wine. Pretty, raw, natural."

On our drive back from Peter's farm, Caleb said to me, "I would rather have a wine transparent to what its intention is." Me, too—and the increasing interest in growing underutilized wine varieties seems to indicate others do, as well. Caleb continued, "So the question becomes, Can we make space for both? For the big, bold red and the crisp, green white? Can we hold them both as our own? Can we be open enough to go the distance?"

After touching the grapevines, meeting Scott and coming to understand the critical need for diversification, I know I want to be that open. When decisions are made in favor of the wine, the grapes and the wine taste different. Peter didn't grow the grapes for money; Scott didn't make the wine for fame. They both did what they did for love and joy. Because it was what they wanted for themselves—and what they wanted to share with us.

Months later, Caleb and I did a virtual tasting. I was nervous, still smarting from my wounded heart and my misidentifications of the aromas in the Le Nez du Vin wine kit. Over Skype, I opened up my bottle of Trousseau Gris as Caleb swirled his in the glass. I told him I hoped to find what I found before at Camino, but worried I wouldn't be able to. He offered a compassionate smile and said, "The specific aromas aren't as important for me as they used to be. I'm looking for the whole of it—the balance, the soul and the heart. It's not about whether I can find jasmine or some other flavor here; it's about what I want or need in the moment."

I couldn't screw this up.

As I swirled and sniffed, Caleb said, "Sometimes you don't know why something tastes good. It just does. What I am looking for is something with balance and good acidity—and something interesting. It's important to remember that things taste different on different days." I asked why. Was it something inherent in the wine or the drinker of the wine? "Both," he said. "It's your mood; it's your intention and the intention of the maker. It's barometric pressure. It's the glass you're drinking the wine in. It's the perfume of the person beside you.

"The intention behind big wines is consistency," Caleb said. "The wines made more naturally can be fickle, petulant and not as consistent. Know what you need; there are wines for different occasions, moods and foods. Not everything you'll smell or taste is necessarily something you'll like. And that's a *good* thing. It means you've ventured outside your comfort zone." Plus, he added, you'll still learn from it.

But this adventurousness can feel challenging if you're sampling a pricey glass or bottle. "It is a risk," Caleb admitted. "People ask, 'If I'm going to pay $12 to $14 for a bottle of wine, why should I pick up something weird?' For me, it's because I know the potential payoff is greater than the risk." Every taste has the potential to bring us pleasure—if we summon the courage to try something new. "And it's never a total loss," Caleb says. "You can always cook with it or find someone else at the table who will finish it off for you."

I started to think of this idea as a metaphor for how I wanted to live my life. I want to cultivate that openness; I want to be a woman who is willing to take risks and make an investment to get closer to what I love. That bottle of Jolie-Laide Trousseau Gris costs $26. Tasting wine feels like a low-stakes way to practice.

We don't need to overthink this experience or make it harder than it is. It's fermented grape juice. Unless we're entering a retronasal competition, all we have to remember is that tasting is simply a way to increase our enjoyment of food and wine. The reason Darin Sukha (the sensory expert I studied with in Trinidad) suggested I smell and eat new things was because it can help us better reveal the flavors of wine, or any food or drink. Every experience helps inform what follows. The best way to begin to appreciate new tastes and smells is to be present to what's before us.

"Pay attention to what you're eating and what you're smelling," Caleb says. "Try to remember not only what you had for lunch today but two weeks ago. Keep doing it. Even write it down, if you like." But don't let memory be your only guide. Learn wine flavors by smelling and eating the foods they remind you of. Combine them in the same ways you find them in the wine. Collectively, these efforts will work together to build up your neural pathways. My Trousseau Gris, for instance, reminded me of fresh peaches, green grass and slate, so I tasted peaches, sniffed grass, and washed and licked rocks to see if I was close. (I was.)

And the process is iterative. Sensory guides, starting with the one on page 344, are almost always in the shape of a circle because smells and tastes exist on a continuum. They work together. See what happens when you pair the wine with fruit, cheese or a meal. Notice what the interaction does to the food and drink: Does the wine amplify or dull the flavor of the food? Do the flavors meld together or bring out new ones? What are those flavors? What do they teach you?

With few exceptions, experts try to avoid descriptors like "good" and "bad." That's because they are absolutes; they don't advance conversation or exploration. If you just need a glass to unwind or accompany your dinner, then absolutes are fine. But if you're on the journey, take your time and find the nuances. Slowing down not only helped me build my sensory library, it transformed my definition of pleasure—and redefined notions of "good" and "bad." It helped me to understand why I loved what I loved—and to feel empowered to allow it to change and to own the experience fully.

A key part of this understanding was fleshing out exactly what informed my choices. When assessing deliciousness, the foundational sensation is smell. Smell is highly personal and deeply intimate. Different people experience the world in different shades of smell: Depending on how active our olfactory receptors are, there is about 30 percent variation between us.[61] This is why some of us love a stinky blue cheese or clove of roasted garlic, while others loathe them. The variance isn't just a reflection of our innate differences; it's also shaped by how often we challenge our olfactory system. By exposing our smell receptors to different scents, we increase the system's plasticity and reshape our smell response.[62]

Yet, despite these variations, nearly every one of us is vulnerable to outside cues. French wine researcher Gil Morrot and his colleagues added an odorless red dye to white wine and presented a glass of white, plus a glass of white tinted red and a glass of *real* red (a blend of Merlot and Cabernet Sauvignon) to 54 students in the wine program at the University of Bordeaux. The students thought they were drinking three wines, not two. They used terms typically used to describe red wine— prune, clove, cherry, chocolate and the like—to describe both the red blend and the fake red. The unadulterated white, however, was described as having typical white wine notes, including honey, lemon, grapefruit and white peach.[63]

Color isn't the only thing that sways us. Frédéric Brochet, one of the co-authors of a 2001 study on perception and wine labels, conducted another experiment and found that, when the labels were switched and wine was presented in an expensive-looking bottle, the test subjects raved about its complexity and balance. The same wine poured from a cheap bottle was described as flat and weak.[64]

Price also influences the very real pleasure we derive from wine. In a joint study conducted by Stanford University and the California Institute of Technology, 20 volunteers were hooked up to a functional magnetic resonance imaging (fMRI) machine to measure blood flow to the brain. The volunteers had to remain completely still while wine was pumped into their mouths. The subjects expressed a consistent preference for the $90 wine over the $5 wine, which—you guessed it—was actually the same wine. And they didn't just say it; the medial orbitofrontal cortex, a part of the brain associated with pleasure, *showed* it.

The perception of a more expensive vintage brought greater pleasure. But the pleasure wasn't intrinsic to the wine. When the wines were presented without prices, the wine that ignited the most pleasure was the cheapest one.[65]

Not only does this reinforce that we should own and love our choices; it also reminds us that these choices aren't fixed. In blind tastings, we try to strip all external cues away in an attempt to focus solely on what's in the glass. But nothing is ever really blind, is it? We carry our stories and perceptions with us, all the way to the wine. No decision is final.

In every conversation I had with Caleb, he talked about intentionality. "Where does the wine come from?" he asked. "Is it from a small vineyard or a big corporation? Are the people who grow the grapes far removed from the final product? This is part of what we taste. There is a different intentionality with small makers. They say, 'I can make the best wine I can and it will change from year to year versus adding or manipulating things to make it homogenous across vintages.'"

This is exactly what Scott Schultz does. "He doesn't manipulate wine," Caleb said. "He intervenes. Scott listens and figures out how to work with the grape. It's a how question: How do I express what *is*, and how do I maximize it? It's honest."

In vino veritas. This is at the heart of biodiversity and the soul of taste. If we want cookie-cutter sameness, then we should keep doing what we are doing. If we want something less uniform and more interesting, then these emerging wines are the ones to seek out.

"Everyone in the industry wants you to drink good wine at whatever comfort level you want to be at," Caleb added. "Most want to give you the best wine they can for your style and price. At ten dollars, you're at a safe price point to get something good. Spain is a great source, France can also do it, but it just might not be with areas or grapes you're familiar with. Be willing to be flexible with the grape you're getting."

The most important factor, Caleb emphasized, is the relationship between you and the wine. "When you buy a wine at your price point over the Internet or in an anonymous setting, you'll get the *price* you want but not necessarily the *wine* you want. Think about what you need and then find a passionate retailer or restaurant that can give you good guidance. If you don't want to spring for a bottle, order a glass or request a sample." Consider the soil, the vines and the grapes. Say yes when a waitperson offers you something new. If she or he doesn't offer, ask. Start where you are and allow yourself to be changed.

"Our threshold for something interesting grows over time," Ann Noble told me, "as we have more experiences and build up our language to describe the experiences we're having." We can, according to Ann, completely transform an experience by changing our beliefs about it.

"If the drink is bitter, turn yourself to wine."[66]

III. TASTING WINE

I put it near my nose and nearly pass out. It smells of old
houses and aged wood and dark secrets, but also of hard,
hot sunshine through ancient shutters and long, wicked
afternoons in a four-poster bed. It's not a wine, it's a life,
right there in the glass.

—NICK HARKAWAY, *THE GONE-AWAY WORLD*

It's time to celebrate the wine! Start with a clear mind and palate. There
is no right or wrong way to experience wine, but there is an *optimal*
way if you're trying to really zero in on what you're imbibing, and that's
to block out as much extraneous stimuli as possible. That said, if barking dogs or screaming kids are part of your life, let them be part of your
experience.

*Uncork or uncap your wine, pour it into a glass or decanter, and let it
breathe.* The goal is to aerate the wine, increasing the amount of surface
area that is exposed to oxygen to allow the volatile aromas to release.
Temperatures and aeration times will vary depending on the wine you're
drinking. People tend to overchill white wine and let red wine get too
warm, but the reference point for room temperature was meant to be that
of a wine cellar. There is a sweet spot in between chilled and warmed
where the volatile aromas will start to release.

Observe how the wine sounds as it hits the glass. Notice if your mouth
begins to salivate as the wine is poured. Fill the glass about one-third full.

Look at the glass. There are a variety of glass shapes and sizes designed to optimize the flavors of wine; try to use a glass that has a slight tulip shape (to retain aromas) and a stem. Hold the glass by its stem so you don't warm your wine past its optimal temperature. Notice the streaks coming down the sides of the glass, called "legs." Wines that have good legs are ones with more alcohol and sugar, which generally means they're bigger, riper and denser than those that don't.

Swirl that deliciousness around to allow for further aeration. As the wine warms slightly and air starts to circulate, the aromas will increase and intensify. As you inhale, open your mouth to allow the scents to move through your nasal passages and olfactory cavity, and think about what you're smelling. The first note might simply be "red wine" or "white wine." Great. Keep inhaling. Assessing the nose of the wine is analogous to sipping: Take a few quick sniffs so you don't overwhelm your system. Be a bloodhound, not a yogi. Pay attention to the memories that are coming up for you. Reach for them—and let them reach for you. And remember: None of this is intuitive. We all learn as we go.

Notice the color of what's in the glass. You can hold it up to the light or against a sheet of white paper to better gauge the tones. Color is an indication of wine processing and vintage. White wines deepen in color as they age; red wines lighten. Notice if or how the color changes as it radiates out from the center of the glass.

Smell again. Every breath is a new opportunity. Have any of the aromas changed or deepened? Wine has various volatile compounds that show up at different times. If what you found before was "wine," see if anything else starts to reveal itself. Close your eyes. Notice how that changes the experience; see if it connects you back to a different memory. Wine is full of relationships—from the growers and makers to the people and places that come to mind as we imbibe.

There's a point when we all experience olfactory adaptation, or fatigue, and lose our ability to distinguish a certain smell (like how we register smells in other people's homes but not our own). The best way to reset is to rest or expose our olfactory system to new smells. Caleb prefers to take a break, but smells the back of his forearm or takes a quick sniff of brewed coffee if he can't rest.

Pick up the glass and take a sip. How does the smell connect to the taste? Sommelier Hande Leimer taught me to gently slurp wine into my mouth (as if you're sucking through a straw, minus the straw) to introduce more air and direct the volatiles toward the retronasal cavity. Play around with this; it will serve you well when sampling any liquid. And also feel free to spit the wine out; the goal is to become enlightened, not drunk.

Let the wine coat your mouth. Swish it around. How does it feel? The "body" of a wine is usually described as light, medium or full, while the texture might be something like creamy, dry (astringent) or effervescent. These components are especially important in pairings, researchers recently learned, because of our mouth's innate tendency to seek balance. The balance of acidity and fat shows up in a Coke and fries, and, of course, wine and cheese.

Take another sip. What's there? What's showing up for you? If it's still "wine" and you're enjoying it, all is well. But stay open and see if anything else turns up. Review the Wine Aroma Wheel found on page 344 for inspiration. Start at the center of the wheel to identify general aromas, and then move out toward the edge to get more specific. Is your wine sweet? Does it smell and taste fruity? Dig into that fruit: Is it citrus or tropical? Berries or cherries? Or is the flavor more of a concentrated, dried fruit? Does it remind you of figs or prunes? Pull entries from the wheel and from your personal sensory library.

Smell and taste are layered and comingled. Once you've got the wine in your mouth, you'll likely detect one smell or flavor at the start and then a few more. Stay with it. See how long the experience lasts, and if anything else turns up in the middle or at the end (in the finish). To clear your palate, chew a small piece of plain bread and rinse with lukewarm water.

Explore different wines in a way that makes sense to you. Try a few of the same varietals from the same vintage made by different producers, or sample different vintages of the same grape made by a single producer. When you feel you understand the wines, consider a blind tasting where all identifying information about the wine is removed. "To take wine into our mouths is to savor a droplet of the river of human history," wrote author Clifton Fadiman.[67]

Imbibe and enjoy.

CHOCOLATE

I. THE BABA

None of it was what I expected: not the smell of damp earth, not the sight of scentless, white blossoms the size of my fingernail or pods the length of my forearm—astonishing, grooved footballs in green, yellow, orange, red and purple growing straight out of tree trunks. Not the sound of the pods thudding to the ground with a swift jab of the *palanca*, a sharp blade on an extended wooden pole that stabs at the small expanse between the pod stem and trunk, or the sound of pods cracking open with the whoosh of a machete or hard strike against a tree. Not the warm, sticky flesh.

I hadn't fully conceived of what a chocolate forest would look like, but it wasn't this, a fantastical Dr. Seuss dream made manifest.

Chocolate comes from cacao (pronounced "kuh-cow," not to be confused with coca, the shrub that becomes cocaine). The fruits are ridged, oblong or round, roughly the size of large cantaloupes or American footballs. Spanning a range of colors, they can be both smooth and mottled, and are attached to tree trunks and thick branches by surprisingly thin stems less than one-fourth of an inch thick. The pods grow vertically out of wide branches and trunks of trees, and their placement looks haphazard, like a botanical game of Pin the Tail on the Donkey.

The pulp inside the fruit—the flesh covering the seeds—is warm, sticky and sweet. However, cacao is harvested not for the flesh, but for the seeds that we mistakenly refer to as beans. Those seeds, once processed, are what become cocoa and chocolate. And chocolate, as we well know, is a miracle substance.

For years, it seemed, I had been nursing a continuously fractured heart—sadness punctuated by moments of such extreme joy, amazement and gratitude that, when taken in aggregate, was something approximating happiness. While I was, as my friends in Singapore would say, "quite okay," my life was undergirded by a deep sense of longing. In those moments, I turned to chocolate, universal salve of the heart.

That is because chocolate, quite literally, heals our hearts. The beans are rich in plant nutrients called flavonoids, compounds found in fruits, vegetables, teas and wines that are beloved for their antioxidant and anti-inflammatory properties.[1] The cocoa flavonoids in chocolate, also known as flavanols, help protect LDL cholesterol molecules from oxygen-related damage, which reduces the chances of coronary heart disease,[2] increases the strength of our blood vessel walls, regulates the constriction of small blood vessels and makes blood platelets less likely to clot in our bloodstream.[3] The more processed and diluted the bean, the fewer the flavanols; the fewer the flavanols, the fewer the health benefits. So chocolate peanut butter pie isn't a heart-healthy option, but a few squares of dark chocolate certainly are.[4]

Cacao also contains theobromine, a molecule similar to caffeine that perks us up,[5] and anandamide, a brain receptor that, not unlike marijuana, brings us back down.[6] It's why Diane Ackerman calls chocolate "a calm stimulation, a culinary oxymoron,"[7] and helps explain why Americans eat just over 9 pounds of chocolate per person per year.[8] The Swiss lead in global consumption, eating nearly 20 pounds (roughly 9 kilograms) of chocolate a year.[9]

Chocolate soothes and ignites—as it has for thousands of years. Ten beans allegedly bought the services of a prostitute for the night for the Nicaro people of Nicaragua, who, along with the Maya Indians and others in Mesoamerica, used the beans as currency.[10] The Aztecs called their bitter blend of cacao, water and corn meal *xocoatl*.[11] It's believed they "ritualistically [ate] cacao off each other's skin during sex, and the Aztec emperor Montezuma ... drank copious quantities of a chocolate-based beverage to enhance his virility."[12] Montezuma allegedly distributed up to 50 golden goblets of liquefied chocolate each day to satisfy his harem of wives.[13]

But it wasn't just a substance of sex. "Both the Mayans and Aztecs believed the cacao bean had magical, or even divine, properties, suitable for use in the most sacred rituals of birth, marriage and death."[14] The Aztecs believed cacao was a gift of Quetzalcoatl, the god of wisdom, while the Maya called the seed "god's sustenance" and associated it with various deities, including the rain god and the gods of death and the underworld.[15] The seeds were a blessing; they were holy.

This delicious blend of sanctity and profanity is why Carolus Linnaeus (the Swedish botanist who developed the classification system for plants and animals) categorized the fruit as *Theobroma cacao*: "food of the gods."

If, two years ago, you had asked me to describe where chocolate comes from, I wouldn't have known. I might have said, with some certainty, that chocolate grows on trees. Maybe with some coaxing I would have envisioned something brown—a leaf or some kind of sap—but not plants thick with leaves bigger than my head. Not fantastical fruits full of creamy, white pulp enrobing 20 to 50 seeds.

In the 16th century, the Spanish invaded the Yucatán Peninsula and Mexico and carried those almond-shaped beans across the Atlantic Ocean back to Europe, as part of the spoils of their conquest.[16] They stripped it of spiritual significance, sweetened it with sugarcane and made it the premier drink of the aristocracy, initially valued for its medicinal purposes but soon appreciated for taste. "Are we shocked to learn that a medicine or drug with supposedly curative powers was converted to recreational use?" anthropologist and co-author of *The True History of Chocolate* Michael Coe asks. "We should not be."[17]

Chocolate even inspired Italian adventurer, author and charmer Giacomo Casanova, who, in his diaries, called chocolate "the elixir of love" and preferred it to champagne, "frequently discussing his habit of consuming cups of chocolate in order to sustain his lustful exploits."[18]

Thankfully, this sexy folklore has now been affirmed by science. Multiple chemicals in chocolate make us feel love—or something like it. This includes the amino acid tryptophan, which is used by our bodies to make serotonin, a neurotransmitter that's connected to heart function and regulates our moods and sexual arousal.[19] It also includes phenylethylamine (PEA), the "love drug," a stimulant in chocolate that's related to amphetamine (speed). It causes our pulse to quicken and is released into our brain when we fall in love.[20]

"But when the rush of love ends," Diane Ackerman explains, "and the brain stops producing PEA, we continue to crave its natural high, its emotional speed. Where can one find lots of this luscious, love-arousing PEA? In chocolate. So it's possible that some people eat chocolate because it reproduces the sense of well-being we enjoy when we're in love."[21] It is also possible that some people seek this same substance to induce a semblance of well-being when they fall *out* of love.

And that's why I had to go to its source. To understand biodiversity, for sure, but also for love—and everything like it. In each of my journeys there was a moment when a single glass of wine or slice of bread changed me. But chocolate was, and is, different. There is no one bar of chocolate that improved my life; they all have.

My first destination was a rundown tropical paradise called Esmeraldas, the northernmost province on the Pacific coast of Ecuador. The country is part of a bean-shaped area in the Amazon jungle that overlaps with Peru and is considered the primary center of origin for cacao.[22] This region is also one of the centers of domestication for the crop— the place where cacao became chocolate. Until recently, the center of domestication designation had gone to Central, not South, America,[23] but a reclassification of the genetic types of cacao has indicated there are *multiple* centers of domestication.[24] (Of course cacao became chocolate in more than one place because, of course, anyone who encountered it would want more.)

Theobroma cacao is part of the mallow (Malvaceae) family of plants that includes cotton, okra, hibiscus and durian. Early designations were based upon only three types of cacao: Criollo ("native"), Forastero ("stranger") and Trinitario ("native of Trinidad"). These three identifications were genetically vague because they were based on appearance, but almost every kind of cacao—including Ecuador's prized crop, Nacional, which is categorized as a Forastero—was slotted into these categories.

The new, more expansive classifications were published in a 2008 study by Juan Carlos Motamayor and a team of geneticists who clustered cacao into 10 genetic groupings that were organized by geographical location or by the traditional variety most represented in that particular cluster (Marañon, Curaray, Criollo, Iquitos, Nanay, Contamana, Amelonado, Purús, Nacional and Guiana).[25] These classifications are based on the actual DNA makeup of the plants, not just appearance.

Mark Christian, chocolate reviewer and director of the Heirloom Cacao Preservation Fund, bluntly (and brilliantly) summed up these distinctions and their expansion. "I am glad the reclassification happened," he told me. "It was a fucking insult to nature to think there were only three varieties of cacao." But three varieties are easier to manage and market than 10, which is why, though incorrect, many chocolate makers still use the original designations.

In wine, the most obvious influences on flavor come from terroir, grape variety and timing (vintage) of the harvest. But those aren't the only influences. Each glass reflects a set of decisions ranging from how the farmer pruned his or her vines to how the winemaker crushed the grapes: With stems? With skins? Crushed by feet or a mechanical press?

In cacao and chocolate, there are also multiple factors at work. (Truthfully, there are *a lot* more factors at work; it's dizzying.) We grow up thinking chocolate is chocolate, but it has complexity and nuance, just like wine. "You think wine is more complex," Brad Kintzer, chief chocolate maker at TCHO, explains, "but it's chocolate. Cocoa has 800 flavor compounds. No other food has as many."

This combination of genetics and origin—and the quest to reveal the best qualities of both—is what led Brad to cacao and chocolate. I had last seen him during my Willy Wonka moment at the TCHO factory in San Francisco. (The factory is now in Berkeley.) Now, Brad and I were reuniting in the only other place on the planet that could bring me the same level of joy as a chocolate factory: the place where chocolate begins.

I met Brad in Esmeraldas. He and his colleagues—Katie Gilmer, former senior sourcing manager, and technical coordinator Aldo Reyes—

were visiting local farms that supply TCHO with cacao. They were also there to help set up TCHO's ninth flavor innovation lab—a joint effort with food purveyor Equal Exchange and the United States Agency for International Development (USAID). The labs were designed to help farmers improve the quality of their cacao and, in doing so, increase their earning potential. I shadowed Brad, Katie and Aldo to understand how flavor could improve these producers' incomes and change the landscape of food and farming.

Before we visited the flavor lab, Brad and I sat down for a round of piña coladas at the open-air restaurant next to my hotel. As he spoke, I pulled multiple tiny umbrellas and garish skewers of fruit off my drink, then started to jot down his story: "I came to cacao through the plant, not the food. When I was at the University of Vermont, where I was studying botany in the late '90s, I'd pretty regularly visit the botanical garden in Montreal. That's where I saw cacao for the first time. After that, when I went on to do fieldwork in Costa Rica, I decided to write a paper on cacao, and it was there that I realized cacao was this amazing way to see the world. I could see how biodiversity manifested—and understand how the world really worked—by looking at this one plant from beginning to end, from the blossom all the way to the end chocolate."

I put down my pen and raised my half-empty glass. In Brad's journey, I could recognize my own.

Later that evening, after multiple rounds of piña coladas, Brad clarified the exact moment cacao took hold of him. "You know that Alfred, Lord Tennyson poem 'Flower in the Crannied Wall'?" I tipsily shook my head. Brad continued, "I was studying botany and feeling overwhelmed by all of it. I wanted a trajectory of just one thing. And then I read that poem." I interrupted and asked if he'd recite it for me. Despite all the background noise—restaurant patrons, festive music, the loud whir of a blender mixing cocktails—the table, all of a sudden, felt quiet.

Flower in the crannied wall,
I pluck you out of the crannies,
I hold you here, root and all, in my hand,
Little flower–but if I could understand

What you are, root and all, and all in all,
I should know what God and man is.

"That poem transformed me," Brad said quietly. "After reading it, I knew I wanted to make *Theobroma cacao* the focus of my botanical work." Brad felt that through cacao and cocoa he'd understand "something beyond our physical human experience." I thought back to the moment of farmers tasting chocolate that he'd described as "sacred." As I eyed the small pile of paper umbrellas and my growing stack of interview notes, I sensed that this journey would be transformative for me, too.

The next morning, the four of us set out for UOPROCAE, a cooperative where farmers in the region work together to process and distribute their harvest, and the place where TCHO was setting up its flavor innovation lab. Farmer cooperatives are important institutions for farmers because, as Cristian Melo, chief research officer at the Direction of Graduate Studies in Ecuador's Equinoctial Technological University, explains, "a single farmer doesn't produce enough beans to get good fermentation. Volume is the key, and you can only reach it if you work with other farmers."

My introduction to UOPROCAE started well before we entered the co-op compound hidden behind steel gates and concrete walls. The air outside was pungent and sour, the alcohol-like scent of fermenting cacao. I instantly fell in love with the smell, one that greeted me at many places I visited as I tried to understand the deep origins of chocolate.

Inside the co-op, two men named Javier (Javier Enrique Valencia Castro and Luiz Javier Meza Cabrerra) started showing me around. Both men oversee the lab. The older Javier (Javier Enrique) was strapping, rugged and handsome—exactly how I envisioned a Latin American cacao farmer would be. The younger Javier was a surprise; he was baby-faced, with gelled hair slicked to perfection. His tight shirt perfectly highlighted the contours of his muscular frame, right down to the nipples.

Around the perimeter was a group of female interns dressed in tight tank tops, tighter jeans and barrettes of flowers secured in their hair, strutting across the plant's concrete drying patio as if they were models on a runway.

Just as I hadn't known what a cacao forest would look and feel like, I was also clueless about the places where seeds become beans. In this compound, the work was grueling but the energy was upbeat; people were laughing, dancing *and* working.

Francisco Bienvenido Peñarrieta, UOPROCAE's young manager, joked around with me all week, trying his best to shift what seemed like my solemn demeanor (which was actually more about me trying to get over jet lag and make sense of cacao than a reflection of my inner state). But when I asked him what he wanted people to know about Ecuadorian cacao, he, too, turned serious. He paused, closed his eyes and, while lightly touching his fingers to his chest, slowly said, "The producer who is growing the cacao is making a big sacrifice. It's a humble, poor person who grows your chocolate but has the healthiest *corazon*, heart."

After greeting the farmers inside the UOPROCAE compound, Brad walked me over to the drying beds and cut open several seeds, which he carefully placed side by side on the bed's wooden ledge. They looked like the empty shells of coquinas, the small clams that butterfly open when emptied. Some of the seeds looked purple, while others oozed amber liquid and were evenly brown. "That's what we want," Brad told me. "That's a properly fermented seed."

A cacao pod is sterile until it's cut open. That's when fermentation begins. Cacao seeds are roughly 50 percent vegetable matter (what we call cocoa solids) and 50 percent fat, or cocoa butter. After the seeds are removed from the pod by hand and separated from their fibrous central cord, they're collected in buckets or bags and transported to the area where they will be fermented.

Depending on the specifications of the cacao (the type, the season and the conditions under which it's being fermented), cacao ferments for anywhere from three to eight days, usually heaped under banana leaves or jute sacks, or enclosed in wooden boxes and trays or wicker baskets. The beans are, in essence, cooking, as the pulp around the seeds gets gobbled up by the yeasts that are present in the air and on the surfaces the pulp comes in contact with. This includes fermentation boxes, banana leaves, the hands of the people pulling the seeds from the rind and the feet of fruit flies and other insects. Microscopic organisms are the catalysts for change: Yeast converts the sugar in the cacao pulp to ethanol, while bacteria generate lactic acid (the acid that sours milk[26]) and acetic acid (the kind that turns grape juice into wine, then vinegar).[27] Proper timing of fermentation is critical. The goal is to ensure the cacao is fully cooked so that the astringency and off-flavors that emerge when lactic and acetic acids are formed are eliminated.

Before fermentation, cacao tastes nothing like chocolate. Yet, as it progresses from seed to bean, it transforms not only in structure, but taste. By about the second or third day of fermentation, acetic acid—plus temperatures that can reach 45 degrees Celsius (113 degrees Fahrenheit)—kill the seed. That death starts the disintegration of the seed's cellular structure and triggers chemical changes that cause the seed's color to deepen and flavor compounds to develop.[28]

Fermentation, explains author Sandor Katz, "unlocks mysterious layers of flavor complexity."[29] Put another way, the death of the seed begets the birth of flavor. Under-fermented beans often taste extremely bitter or astringent, while over-fermented beans can taste putrid or hammy. "You can have good genetics," Brad explains, "but if you don't process it well, you're not going to get good chocolate."

Until this exploration, I didn't know chocolate was a fermented food. But nearly every moment at the co-op reminded me that fermentation is transformation. Over three sweltering afternoons, I watched the group of fashionable young interns—Laila, Mireya, Karla and Fanny—using more strength than I thought possible, push fermenting beans from one wooden box to another, then scoop them from the boxes into wheelbarrows, lay them out on the concrete patio to dry and sweep them into small, neat piles to prepare for packing. The effort it took to convert the fresh seed into a flavorful bean was back-breaking, but these young women seemed to glide from one place to another, their glamor still intact.

Not every visit to a cacao farm or processing center was as spirited or meaningful as this one, but it was an important introduction. It set the tone for what was possible. The work is hard and tedious, yes, but this isn't a singular story of misery. Poverty and impoverishment are not the same. In every community I met, with every food and drink I explored, people reached for joy.

The beans the girls processed were initially damp, a rich yellowish-brown color you'd recognize as chocolate. But after being raked flat onto drying beds, they turned an ashy brown. From afar, they looked like pebbles. Close-up, they became little jewels, a fruit I fell for the way Brad had fallen—root and all.

The next morning, Brad, Katie, Aldo, Luiz Javier and I headed off to Atacames, a small town about an hour's drive from the cooperative. We were going to visit Alberto Bautista's farm, which he runs with his wife, Alexandra Arces, a part-time manicurist whose smile fills up her entire face and whose nails were adorned with silver and emerald flower appliqués. A former shrimp fisherman and cement factory worker, Alberto purchased the farm five years ago and now tends to it 12 hours a day. He keeps meticulous records and knows his farm intimately. He's also gorgeous—like, supermodel gorgeous—and deeply in love with Alexandra and their three children, Eri, Anai and Emile (which, of course, makes him all the more endearing). His family is his first love, and his farm is his second.

Like most vibrant cacao farms, Alberto's looks and feels like a forest. It's overflowing with vegetation (including cacao leaves up to 2 feet long), steamy with heat and moisture, and swarming with mosquitoes and midges (tiny, biting flies no bigger than a pinhead that are also known as "no-see-ums"). What looks, at first, like a mess of a jungle is actually a planned ecosystem. Alberto's farm is a field laboratory—a place where he can observe firsthand how different varieties of Nacional cacao and other crops he grows alongside the future chocolate respond to everything Mother Nature sends their way.

Midges, the primary pollinator of cacao's tiny orchid-like flowers, are what keep the ecosystem flourishing. They, along with mosquitoes, were relentless, buzzing near my eyes, biting through my clothes, attaching themselves to any spot of skin that hadn't been sufficiently doused in DEET. Relentless—and essential. Essential because, as biologist Sara Lewis explained in her TED talk on the magic of fireflies, "every time a species is lost, it's like extinguishing a room full of candles one by one. You might not notice when the first few flames flicker out, but in the end, you're left sitting in darkness."[30]

The same can be said for pollinators. In the case of midges, this is the result of their loss of habitat. Cacao is increasingly grown on open plantations rather than in dense tropical rainforests. According to the U.S. National Park Service, "within commercial plantations, the time of peak flower abundance is out of sync with the peaking of midge populations. To make matters even less attractive to the midges, wild cacao flowers have more than 75 distinct aroma ingredients, while cultivated cacao flowers have only a few."[31] This is why, though plantation cacao flowers are abundant, "only three out of 1000 flowers become pollinated and produce seed pods."[32] The odds are significantly better—but still slim—in more diverse growing environments where 1 to 5 percent of flowers are successfully pollinated.[33] Once pollinated, the pod takes about six months to fully mature.

Every pod is precious.

Cacao is a crop that has to be continuously grown in order to be sustained, both in *ex situ* collections as well as *in situ* on farms or in the wild. The largest and most diverse *ex situ* collection is located at the International Cocoa Genebank in Trinidad—3,337 kilometers (about 2,000 miles) from Alberto's farm. It is one of more than 35 collections around the world that maintain more than 24,000 accessions (or samples) of cacao. The collection is expansive, but there is no way it can contain every known variety.

We save what we think we need. For scientists, that has meant plant materials that hold the possibility of increased productivity and tolerance to environmental factors—what can help breed what are collectively known as "improved" varieties. But, until recently, "improved" has never meant improved *flavor*. Ed Seguine, a former chocolate research fellow at Mars and founder of Seguine Cacao Cocoa and Chocolate Advisors, and Darin Sukha, research fellow at the Cocoa Research Centre that oversees the International Cocoa Genebank, are at the forefront of a push within cacao conservation to include flavor as one of the parameters for selecting what's saved in genebanks and shared with farmers. "Diversity means different things to different people," Ed says. "If we recognize it, we have a good chance of preserving it."

What's also important to note is that not every plant within the scope of what's been collected *ex situ* thrives. As Darin explained when he gave me a tour of the cacao genebank, some varieties don't bear fruit in Trinidad, as they would in their home countries. Others, meanwhile, show diminished levels of resistance to disease and tolerance for heat. "That's why we need farmers to grow diverse varieties, too."

Yet, farmers like Alberto and his family are part of a shrinking group committed to growing prized Nacional over another Ecuadorian cacao variety called CCN-51. CCN-51 is a hybrid with big pods, big seeds and greater tolerance to disease—but none of the flavor qualities of Nacional. This is because, as plant geneticist Michel Boccara explained during my visit to the Cocoa Research Centre, "Nacional is a cluster of varieties from low altitudes that have been selected over time. CCN-51 is a cross of plant material from the Amazon, plus the highly productive hybrid IMC-67 and the clone ICS-95—there's little diversity there."

As we ventured deeper into Alberto's farm, the incessant buzzing of mosquitoes and midges was punctuated by a much more pleasurable sound: the thud of pods falling to the ground. Ahead of us, Alberto whacked the pods off the lower branches and trunks of cacao trees with a machete. For the pods on the taller branches, he chopped them off with a *palanca*, a hoe-like tool roughly 4 feet in length with a sharp, flat blade. Although heavy, the pods are suspended from their branches by impossibly small and narrow stalks, which are, at most, about one-half inch long and one-quarter inch thick. Expert cacao harvesters, like Alberto, can sever the stem in one motion, as easily as you or I can chop an apple.

One of the first rites of passage when visiting a cacao farm is having a go at harvesting. Alberto handed me the *palanca*, which ran the length of my body. I protested, explaining I was way too short. Alberto then passed it to Alexandra, who is exactly my height. She felled the pod with one quick stab and smiled encouragingly as she passed the *palanca* back to me. I relented and jabbed at the fruit in a motion I imagined to be a close approximation of hers. It felt like I was jousting with a balancing beam. I massacred the pod.

Once people far more expert than I harvest the pods, they split them open with a strong bash against a tree or, ideally, two swift and precise cuts to the ends of the fruit and one slice down its length. I can still conjure the sound of the machete whooshing through the air for the end cuts and digging through the thick rind of the pod to reveal the pulp. It's what I remembered long after I left the lush forest: metal sinking into flesh.

In that first moment, every sight, touch and sound led up to the anticipation of that flesh—the smell and taste of the sweet, sticky, hot, wet pulp of the fruit that was nothing like what I expected. The ivory pulp, or mucilage—known in Spanish as *baba* (literally "slime")—varied not only in taste, but also in intensity. Some pieces of fruit were juicy and thick; others nearly dry with only a thin layer of pulp surrounding the dark seed.

"Go for the middle," Brad advised, while Javier gestured for me to dig inside. There was no neat way to do it. I tentatively dug my DEET-covered hand into the gooey pod and started to pull apart fruit that was held together by the thick membrane I later learned was the plant's placenta. I tilted my head back slightly and dropped the fleshy seeds, one by one, into my mouth. The group watched as my eyes widened and my mouth burst into a smile. It was ... astonishing. I had expected something that tasted like chocolate. Not this: not lemonade and honey-dew, not custard apple and peanut brittle. Greedily, I reached for more and more. Each pod was different: some puckeringly tart, some sugar-sweet, some tart and sweet simultaneously. There were so many tastes, I doubted I'd ever be sated. These were the tastes of biodiversity.

Brad smiled and nodded. "It's a big orgy in here," he said. "That's the promiscuous nature of the cacao tree. It's hard to control without a monocrop." This sums up one of the greatest challenges—and greatest gifts—of agricultural biodiversity: its serendipity. Without intervention, each harvest is a surprise.

There are a few ways cacao can be grown: planted from seeds, grafted onto preexisting saplings, or cloned. Planting from seeds is the biggest wild card because of the nature of the plant. As with apples, cacao doesn't reproduce true to type; every offspring is different. The two ways to ensure consistency are 1) to cut off a portion of the plant and graft (fuse) it onto an existing plant, which is what was done with European grapevines when their roots became infested with phylloxera, and 2) to clone the plant, which involves mass producing plant tissue in a laboratory and is beyond the scope of most farmers growing cacao.

Alberto utilizes both serendipity and precision, planting some trees from seed and also grafting cuttings from his heartiest plants. These plants are considered the original cacao of Ecuador and are collectively known as Nacional. The *baba* I tasted ranged in flavors from peanut to honeydew to limeade. Once processed, they become cocoa that has a mosaic of tastes and aromas, which Brad described as "a balance of floral and nutty with a rich cocoa depth." This same kind of tapestry effect happens in other countries of origin known for their fine flavor. "There is so much complexity in the world of cocoa," Brad explained. "It's one of the greatest joys of the work. It makes the world small and beautiful and connected."

Cacao can only grow in a narrow belt 20 degrees north or south of the equator.[34] In the Ivory Coast and Ghana, the majority of cacao is produced by farmers working on family-owned plots of about 5 to 10 acres.[35] The global chocolate confectionary market is worth around $80 billion,[36] but, according to the Fairtrade Foundation, "growers in West Africa ... are likely to receive just 3.5 to 6.4 percent of the final value of a chocolate bar, depending on the percentage of cocoa content. This is compared with 16 percent in the late 1980s."[37]

Cacao farmers like Alberto and members of UOPROCAE sweat and toil, absorb the risk for the possibility of crops failing, get paid far less than you'd imagine and rarely, if ever, taste the fruits of their labor. For most producers, cacao is an investment, not a treat.

"Cacao is grown as a cash crop for income generation," Brigitte Laliberté, the scientific advisor on cacao conservation at the nongovernmental organization Bioversity International, explains. "Farmers don't consume it or see the end result of their efforts, so the incentive to produce something of quality is very different than crops that are processed and consumed by the people who grow them."

Although cacao is worth more when it's properly fermented and dried, most isn't. And, honestly, for a 99-cent candy bar, it really doesn't matter. Most commercial cacao is blended together, either to get the volume required for proper fermentation or because manufacturers and consumers don't require differentiation. Put another way, the farmer who carefully tends to his richly diverse crop, separates out the moldy beans, gets rid of the insects and carefully dries and transports those beans, gets paid the same amount of money as the person who simply grew the crop as just another commodity.

Most cacao farmers get paid by weight, rather than for the beans' flavor qualities, so they can't afford to invest extra time and energy in quality control. That's why when Brad's colleague Aldo reached into a bag of freshly harvested cacao at the UOPROCAE co-op, he also pulled out bits of tree bark, cotton rags and thick cacao rinds. Because there's little financial incentive to improve flavor, process trumps product—even at UOPROCAE. Once those masses of beans are blended together and additional ingredients are added, a multitude of sins can be hidden. It's the equivalent of drowning coffee in milk, whipped cream and caramel; you really don't know what the foundational ingredient (coffee beans) tastes like.

While showing me around UOPROCAE, Francisco underscored the harsh realities of the cacao industry. When he was 17, he told me, he left his office one day with cacao selling for about $15 or $20 per quintal (100 pounds). "When I came back later that same day, the price had gone down by $1.40. One day you'd get one price, another day you'd get another. Farmers were powerless."

Unfortunately, most still are, explained Jean-Marc Anga, executive director of the International Cocoa Organization (ICCO). "They are the weakest link in the value chain," he told me. "Everyone seems to make more profit but farmers. We should stop pretending we all care about the farmer. We all care about profitability."

Efforts like the flavor innovation lab may shift this balance. The full flavor lab installation—which costs less than $10,000 and is, as described by Brad, "scrappy, not crappy"—includes spice mills, a home coffee roaster, a hand-cranked breaker to split the beans, a tempering stone, chocolate molds, white plastic buckets and a Conair hair dryer. "All this," Brad explained to UOPROCAE members, "will allow you to make your own chocolate from your own cacao—and really see the impact of your work." The goal is not for farmers to produce high-quality chocolate but, rather, to understand and taste for themselves how much their harvests affect the quality of the end product.

There are other elements of the lab (most notably temperature probes and computer software) that are a lot more high-tech than a few plastic buckets and a Conair, but they're all set up with the intention of helping farmers grow and process high-quality cacao that premium chocolate makers want. The tools introduced to UOPROCAE will allow Francisco and his fellow farmers to focus on flavor, ideally increasing manufacturers' (and consumers') commitment to quality, as well as quantity. "Without the sample labs," explains food journalist Corby Kummer, "[TCHO] was in the same position as other chocolate makers: by the time the fermented beans arrived, it was too late to make any changes. The only choices were to place an order or wait till the next harvest. Now, a farmer can make adjustments that will ensure a sale."[38]

But what, exactly, are we getting for our money? There's a qualitative difference between what's mass-produced and what's produced in smaller batches, otherwise known as "craft," "fine" or "specialty" chocolate. Typically, big companies aren't interested in exploring the diversity of taste because brands demand consistency and, according to Brad, no one's going to pay more than $1 for a bag of M&M's. "Nothing changes until we're willing to pay more for a unique flavor experience," he said. "If we customers demand it, they will provide it."

But unlike other industries, such as coffee and beer, there isn't a single definition of what fine chocolate is. The distinction is loosely focused on regions and varieties of beans, but craft makers are now creating delicious chocolate with different beans from all kinds of places. "The combination of plant genetics, location, fermentation, drying, and blending into chocolate and drawing out the flavors is what distinguishes one chocolate from another," Brad explained.

Brigitte Laliberté and other conservationists believe the distinction between bulk and fine chocolate shouldn't be attributed to specific countries but, instead, to specific uses of cocoa in the end product. "If we keep talking about these two worlds of bulk and specialty chocolate, then we will never bring quality into bulk and recognize the quality and flavors that can be grown in different countries." And the point of reference for researchers and chocolate manufacturers exploring sensory qualities of chocolate, Brigitte reminds me, are so-called bulk beans from Ghana, known for their dominant cocoa flavor. "The cacao grown in Ghana is what we come back to," Brigitte says. "It has the flavor we love—the smell and taste of chocolate."

Despite this Ghanian baseline, these designations do matter because beans from certain origins fetch a higher price. The starting price is set through the ICCO and the International Cocoa Agreement, which stipulates that a council of experts decides what countries grow fine-flavor cacao as a percentage of overall production. Currently, nine countries, including Bolivia, Costa Rica, Jamaica, Mexico and Madagascar, bear the designation of growing 100 percent "fine and flavour" cacao.[39] As such, their beans command a premium.

Experts involved in craft chocolate don't necessarily see these bulk/fine distinctions in the same way as scientists or the ICCO, but what everyone agrees on is that some beans have more depth than others, some are better processed after harvest than others, and diversity is essential in order to keep the crop growing in perpetuity.

Every person involved in chocolate knows if we don't make conservation a priority, cacao won't survive.

The support of diversity is like a pyramid. Craft chocolate (like specialty coffee) makes up a small, but growing, percentage of what's manufactured and sold. It has the potential to set the tone for the whole industry and ensure that, in the future, we'll have access to beans that aren't just prolific but delicious. So what separates a 99-cent candy bar from, say, a $6 or $18 bar of chocolate? The first distinction lies in the term "candy." Take a look at your favorite candy bar. Mine is Twix, but, for comparison's sake, let's look at chocolate bars from Hershey's (which sell for about $1), TCHO ($6) and Rogue Chocolatier—whose maker, Colin Gasko, has consistently been named one of the best craft chocolate makers in America. His chocolate bars retail for between $13 and $18.

A standard chocolate bar is typically made up of cocoa mass, sweetener and an emulsifier to improve its texture and consistency. The cocoa percentage listed on a bar indicates the portion of ingredients that come from the bean: cocoa solids and cocoa butter. So, if a bar contains 70 percent cocoa mass, the remaining 30 percent of the bar is made up of other ingredients, most likely sugar or some other sweetener, and an emulsifier. Milk chocolate usually contains between 10 percent and 30 percent cocoa mass, plus milk solids and other ingredients.[40]

In the United States, a bar has to contain at least 10 percent cocoa mass to be called chocolate; otherwise, it's considered confection.[41] A plain Hershey's bar contains 11 percent cocoa mass,[42] while the other 89 percent is made up of sugar, milk and milk fat, two different kinds of emulsifiers, artificial vanilla and other unnamed artificial flavors.[43] In comparison, the TCHO "Classic" bar contains 39 percent cocoa mass, with sugar, whole and nonfat milk powders, an emulsifier and vanilla beans constituting the remaining 61 percent.[44] Rogue Chocolatier's Balao bar is 75 percent cocoa mass and 25 percent sugar.

Rogue doesn't add any additional ingredients because Colin prefers to let the beans represent themselves. His chocolates are single-origin, which means they come from a specific place that has its own history and culture. While there are many efforts underway to process cacao in small batches, in order to properly ferment the crop, you need quantity. That's why—unless a farm is big enough to ferment a substantial amount of cacao—the single-origin designation usually refers to a region or a cooperative group of farmers, like UOPROCAE.

What TCHO and Rogue both try to do, albeit in different ways, is introduce chocolate eaters to a more expansive world of flavor. TCHO's bars are blends of different types of beans, intended to draw out a handful of flavor profiles as a solid introduction to the broader world of chocolate, flavors that they categorize into their own chocolate miniwheel as nutty, earthy, floral, fruity, chocolatey and bright (meaning a bar with greater fruit acidity).[45]

When a chocolate manufacturer makes a bar with less cocoa, there's more room for other kinds of flavors to shine through. That single ingredient (cocoa mass) becomes less important. It also means the quality of the beans can vary and that the quality of flavor—or lack thereof—won't readily be noticed. Chocolate makers who want you to notice the beans and their origins, on the other hand, will work to highlight these differences in their bars.

But those smaller makers—and, by extension, smaller growers—are the ones who face the largest struggle. "Historically, it's been a volume business," explains farmer and chocolate maker Seneca Klassen of Lonohana Estate Chocolate in Hawaii. Seneca is one of the few people both growing and making chocolate in the United States; he oversees the process from cacao pod to the end bar. "We haven't yet created an effective model for small producers or makers," he says. "Small is beautiful, but the commodity system is set up to extract maximum value from the developing world, which results in permanent impoverishment of farmers and maximum profit for manufacturers." Yet, he remains hopeful. "We're trying to unlearn 150 years of industrialization to reward small farming and small manufacture, and it's happening by degrees, in little pulses."

Colin is a little less optimistic. Even though this is his life's work, he is adamant that "craft won't save chocolate." He maintains: "A highly specialized product isn't the solution. We need to have a fundamental shift in how we approach chocolate in general. I have a problem with the idea that the only people who can help the system function in an equitable way and increase value should come from the narrow specialty market."

Colin doesn't want to create a chocolate that only few can afford, but one of the greatest challenges is that cost efficiencies are dependent on scale. "There's a huge chasm between big and small," Colin explains. "With a few exceptions, small makers do everything by hand, not machine. You'd think with a chocolate selling for $13 we're making money, but no one is."

I don't eat expensive chocolate to be fancy or waste money; I eat it because I want to support the chocolate makers and farmers dedicated to sustaining diverse and delicious chocolate. I eat it because the best versions of this are like nothing else. And I eat it because I don't want my joy to come at the expense of someone else's misery. Although rare in Ecuador, cacao farming in West Africa often includes servitude and child labor.[46] As Robbie Stout, co-founder of Ritual Chocolate, explains, "If you're buying cheap chocolate, it's a sure sign someone is getting screwed."

And, in all likelihood, those who *are* getting screwed have the most to lose; they are the ones operating on the slimmest of margins. Even after accounting for as many factors as possible, the journey from blossom to bar is full of risk. An infestation of disease could decimate the crop before it's harvested—or a spate of insects could contaminate the beans on the way to their Willy Wonka destination. Dedication and care are at odds with a larger system that rewards quantity above all else; quality, in an industrialized context, doesn't put food on the table.

Rogue is a company of two people; Colin selects and orders the beans, hand-sorts them when they come in and oversees every aspect of manufacturing—from roasting to wrapping. As a small producer, Colin wants to have a special, differentiated product, which means he's sometimes forced to buy smaller quantities of beans that have to be sent via air. (Delivery in an airplane is much more expensive than shipping full ocean freight containers.) "And then there's a 30 percent loss in the bean after it's roasted and the shells are removed. So I'm paying to ship across the world shells that I throw away. It really annoys me for financial—and environmental—reasons."

And it also explains why his margins are low. But Colin keeps at this, in part, because he knows "there's a greater likelihood of reduction in genetic diversity if we don't value a differentiated, specialty market. My value isn't in purchasing power; it's in showing the potential of what good cacao processed in the right way can taste like."

I spoke to Colin and Seneca in the long shadow of news stories that insisted the only way we'll have chocolate in the future is to settle for varieties with subpar taste, including a variety that Ed Seguine, one of the most respected researchers and consultants in the industry, says tastes like "acidic dirt."[47] But this is only our choco-future if *we* choose it. Ed has spent over three decades working with cacao and chocolate and knows that care follows attention: If we start to recognize the diverse aromas and tastes in chocolate, then we'll understand why they're worth saving. Plus, that diversity doesn't stand alone; when we have diversity in taste, we also get diversity in variety and other important areas.

Chloé Doutre-Roussel, chocolate expert and author of *The Chocolate Connoisseur*, underscores, "If we want diversity, we have to pay for that uniqueness and be willing to tolerate some unpredictability; it is a plant that changes. We should accept a range of flavors related to the fact that cacao is a crop. But we should not accept bad work from the farmer— cacao that has been poorly fermented or badly dried or stored—nor the chocolate maker—burned beans, poor texture and so on."

Brad feels the power is, ultimately, in our hands: "Either the big guys include more diversity or the small guys continue to grow. That's up to us, the consumers." We reshape the system through *in vivo* conservation— saving chocolate by eating it—transforming what's grown and sold through our tastes and our choices (in chocolate and other foods).

The first Rogue chocolate I tried was Colin's $13 Balao bar, which he named after the beans from a farm in Balao, a town in southern Ecuador. It smelled like honey and tasted sweet but earthy, rich with spicy flowers. I didn't know it was humanly possible to eat chocolate slowly, but I learned that when it's complex, it's the only way—akin to reading a rich novel over skimming the morning paper. You want to linger and take it all in. A story with depth fills you up in ways you couldn't know until you encounter it. Before I tried the Balao bar, $13 seemed like a steep investment, but later it became my affordable luxury. By "luxury," I mean it isn't candy. It's not a 99-cent confection I picked up in the checkout line at a grocery store. It's something I savor over *days*, not minutes. And it's available to anyone willing to take the risk and make the investment.

Chocolate is currently dominated by a small group of multinational companies, including Mars Incorporated and Mondelēz International.[48] They buy most of the world's cacao and, as a result, dictate what's grown, thereby defining the parameters of what chocolate tastes like. Every bite of chocolate consumed within—or outside—of those parameters has the power to either perpetuate or transform what we eat. If we stay on this same course, we *will* end up with acidic dirt.

Through years of study, researcher Darin Sukha learned cocoa flavors are the result of many factors. Floral flavors, he discovered, are mainly a reflection of genetics (what's inherent in the seed), while fruity flavors are largely the result of terroir, the taste of place.[49] But all of these attributes can be pushed and pulled depending on how the bean becomes a bar and how much care goes into the process.

It takes Alberto four years to grow plants from seed and two years from grafts. The plants bear fruit for about 35 to 40 years, and then yields start to taper off. Every time yields diminish or plants die, he and his fellow farmers are faced with a choice of growing traditional varieties, such as Nacional, or replacing them with high-yielding hybrids or clones. CCN-51 used to be the default choice, but that's now changing. Alberto's modest yields of Nacional can grow significantly when coupled with modernized growing techniques (such as pruning trees, spacing plants for increased airflow and using fertilizers), and he is doing his best to stick with what he has. "This cacao," he says, "is the blood of the earth." If he removes the dwindling amounts of Nacional that remain, it will likely disappear.

Although Alberto makes the decision sound easy, I know it isn't. He is young and hopeful, but many farmers involved with cacao, especially those in Ivory Coast and Ghana, are edging closer to retirement. Younger generations of farmers taking their place want to grow what's lucrative, a cash crop that will give them an easier life and steadier income.

The numbers in Alberto's meticulous ledgers only add up because he raises livestock and grows a number of different crops. But he remains committed to the value of traditional cacao: "I grow Nacional because it is the cacao of my people *and* because it is what the market wants. There isn't any special market for CCN-51. It is ordinary, and it makes me compete with farmers in West Africa who sell their cacao for a lot less."

Italian poet F. T. Marinetti wrote, "What we think or dream or do is determined by what we eat and what we drink."[50] Over its history, chocolate has been defined as many things: an aphrodisiac, a medicine, a holy food, a balm people rely on to heal a broken heart and a way to show love. These small morsels—and the choices embedded within them—help us manifest the changes we want to see.

II. THE BAR

As I finally got my fill of *baba*, Brad lined up pods and sliced each one open, assessing the seeds. "They're a general indicator of the genetic mix of the farm," he said as he scraped off the pulp and cut a few seeds from each pod. Shaped liked flattened almonds, they revealed shades of pale white and violet and the graceful swirl of the cotyledon, the part of the seed's embryo that later becomes the nib. These seeds were a mix of what were originally identified as Criollo and Forastero beans, ones native to Esmeraldas.

After sucking off the flesh, I gently bit into the soft seeds, reasoning that if the pulp didn't hold the taste of chocolate, the seeds surely would. But the astringent seeds that buried themselves in my molars held no promise of chocolate, no hint of what the beans would become. The flavors I sought would only emerge once the seeds were fermented and dried, and reach their full potential once roasted. I stuck my dirty fingers into my mouth and dug the remnants out of my teeth. "It's chemical warfare," Brad explained. "The seed was designed by nature not to be eaten."

That makes perfect sense. Cacao wants to live long and prosper— just like us. From an evolutionary perspective, bitterness is an indication of poison, which is how (and why) cacao proliferated: Animals would suck the pulp and spit out the seed. This is what author Michael Pollan described as "the botany of desire."[51] A plant makes its nectar smell sweet or its fruit taste delicious or intoxicating to lure us in. The plants make us want them because they need us as much as we need them. We're interdependent; as humans domesticated plants, they do-mesticated *us*.

In the case of cacao, the plants offer up pretty little flowers in shades of white and pink whose shape is the perfect landing pad for midges. They attract the pollinators by producing a sugary liquid from tiny glands on their stems and the base of their flowers.[52] Then, blossoms grow into pods whose festive shapes and colors attract the attention of humans and animals who are seduced by the luscious pulp inside the enticing colorful rind.

The bitterness of the seed, however, guarantees the love affair is fleet-ing. Whoever—or whatever—cracks open the pod will savor the fruit but discard the seed. It's the way cacao ensured it would be propagated: The discarded seed could grow again. But unbeknownst to the plant, the seed has become its salvation—a substance that, once roasted, becomes even more coveted than the pulp. The seeds aren't discarded by humans; they are the reason we plant more cacao. The bitter seeds beget the sweet chocolate.

While the path from grapevine to glass, and the journey from amber waves of grain to loaf of bread feels familiar, the passage from Seuss-like footballs to creamy comfort food remains mostly a mystery. It's winding and full of many steps, each of which can dramatically impact the flavor of the finished product. And here's how it happens.

It takes about 400 beans, or approximately 11 pods, to make 1 pound of chocolate.[53] Yet, every harvest produces a different yield, depending on a number of factors, including the ages of trees, and how they are pruned and maintained. Adding to the complexity are fermentation protocols that can—and should—change depending on whether the harvest season is wet or dry. (Most cacao-producing countries have one main harvest and another smaller one.)

The seeds we call beans are roasted, shelled (or "winnowed") and cracked to expose the cocoa nib. Next, the nibs are ground into a paste known as cocoa liquor, which can then be directly processed into choco-late or pressed into a powdery disc, or cake, from which the cocoa butter has been separated out. From there, the "presscake" is turned into cocoa powder or, later, recombined with cocoa butter or other vegetable fats to make chocolate. (Vegetable oils, like palm oil, render a much lower-quality chocolate.)

The bean is 54 percent fat and is stable with a long shelf life, a fat that's solid at room temperature but starts to melt in our mouths or under our touch. Its stability means it's coveted not only in chocolate but also as an ingredient in medical and beauty products. The butter itself has a hint of cocoa flavor and is the only part of the bean used to make so-called "white" chocolate.

If the butter was separated out from the cocoa solids when it was made into a presscake, it is added back in during this stage of the chocolate-making process, often supplemented with an extra portion of cocoa butter because fat—glorious fat—makes the chocolate creamier and, as the carrier of cocoa's aroma compounds, more flavorful.

The resulting mass (with any added ingredients—usually sugar and possibly an emulsifier to make the chocolate smoother) is now the texture of wet sand. As it's milled through a series of rollers and kneaded (or "conched"), the size of the chocolate particles gets smaller, and the chocolate gets smoother and silkier. Conching refines texture, rounds out flavors and, with heat and time, helps release the fat from the cocoa nibs.

This fat is precious. Mark Christian writes on his website, the C-Spot, "Cocoa butter has been used for centuries as soap, an emollient, and an added fat. The Jesuits, huge chocolate drinkers during the European invasion of the New World, were first attracted by extracted cocoa butter's medicinal healing properties. They established infirmaries and pharmacies, giving rise to the saying: 'where there's a church, there's chocolate.' "[54]

This attraction to fat, however, goes far beyond medicinal properties, all the way back to our evolution. "We are hardwired from birth to like certain tastes and smells, most notably those of sugar and fat," says neurobiologist David J. Linden. "Large, fast-rising pleasure signals are the most rewarding and most addictive." Sugary and energy-dense fatty foods, like chocolate, light up the pleasure circuit in our brains, often influencing us to eat beyond satiation. In an experiment with rats, researchers found that, even after their fill of lab chow, rodents will "readily consume more food if it is sweet or fatty." They continued, "Then again, we didn't really need an experiment with rats to tell us this: Everyone has had an experience of feeling full at the end of a meal but still having 'room for dessert.' "[55] We also tend to eat more when we don't have to invest too much energy in chewing or swallowing[56]—the smoother the food, the better.

Particles of food (and other materials) are measured in microns, short for "micrometers." One micron equals 1/25,000 of an inch. For perspective, a red blood cell is 8 microns and the diameter of a human hair is 50; most humans can't see anything smaller than 40 microns unless it's magnified.[57]

But our tongues, those ever-sensitive conglomerations of muscles, can detect differences in particle sizes under 20 microns. Through refining (milling and conching), a chocolate that starts out with particles of about 100 to 150 microns is, ideally, reduced down to about 15 to 20 (which results in a smoother texture). A chocolate with particles over 30 microns will register on our tongues as gritty.[58] That sensation influences the entire experience of flavor. "The whole process of making chocolate is to break down particle size and expose flavor," explains chocolate maker Matthew Escalante. "Every step of processing changes the possibilities."

The next stage is tempering: realigning the fat crystals in cocoa butter through a controlled combination of heating and cooling to increase the chocolate's sheen and intensify its snap. Tempering is tricky; if the cocoa mass isn't tempered properly, it has a greater chance of getting fat bloom, the whitish coating or splotches caused by the separation of cocoa solids and cocoa butter. After tempering comes the sublime moment when the molten liquid is poured into molds, cooled and—finally—packaged for consumption. *Finally.*

The process of making chocolate requires the same core skill needed in tasting it: patience. It takes time to create good things—and to taste them. As the heat of our bodies melts the chocolate, the flavor compounds suspended in fat (cocoa butter) are slowly revealed, making the delivery system and the resulting sensory experience like no other. If we want to taste the full expanse of flavors in chocolate, we have to slow down and let it melt in our mouths and reveal itself fully. "Chocolate is one of the most complex foods we know," Brad explained, paraphrasing author Rowan Jacobsen. "It can be bitter, sweet, fruity and nutty all at once. It takes a vast library of smells and tastes, and blends it into one revelatory package."[59]

In order to fully access that revelatory package, we need every one of our senses and sense organs, but especially the nose and tongue. As we register taste, texture and temperature, our mouths unlock suspended flavors and turn our food into a chewed-up mass, known as a bolus. Although there isn't much chewing involved with chocolate, when there is work to be done, the eight muscles that make up the tongue spring into action: "Unlike other muscles, such as the bicep, tongue muscles don't develop around a supporting bone. Rather, they intertwine to create a flexible matrix, forming ... a structure [that's] similar to an octopus's tentacles or an elephant's trunk."[60]

The four superficial (extrinsic) muscles attach to bone and change the position of the tongue while the four unattached (intrinsic) muscles, which change the shape of the tongue, are used to push food toward our back molars to help turn it into a bolus. In doing so, the volatile aromas locked up in foods are released, making every bite more flavorful.

The fleshy bumps that dot our tongues are often (mistakenly) called taste buds. Those are actually taste papillae. Taste buds are embedded *within* the various types of papillae that are found on our tongues, throughout the oral cavity, including the nose and throat, and even in the intestines and the passageway to the lungs.[61] They, along with smell (olfactory) receptors, are "the fastest growing and most rapidly regenerating cells in the body. Taste buds regenerate completely in a 24-hour period [and] the entire bud is replaced by new cells on a daily basis."[62] This remarkable regeneration slows down as we age and tapers off in adulthood to about 5,000 working taste buds.

Each taste bud contains up to 50 sensory cells,[63] which are coded to recognize five basic tastes—sweet, sour, salty, bitter and umami—plus fat, which is referred to as our "sixth taste."[64] Fat is something we've only begun to understand, but its presence—or perceived presence—affects our understanding of food. For example, in a study from the University of South Florida, 65 students were given chocolates identical in ingredients and calories, but presented in hard and soft forms. The participants greatly overestimated the "perceived oral fattiness" and calories in the chocolate when it was soft as opposed to when it was hard—from 65 calories up to 101 calories.[65]

Sensitivity to fat differs from person to person.[66] Researchers used to think taste buds on different parts of the tongue mapped to particular tastes ("bitter in the back, sweet in the front"), but we now know every one of these sensory, or receptor, cells responds in varying degrees to all tastes by detecting chemicals in our foods and drinks through hairlike projections called microvilli.[67] These microvilli connect with molecules of food that are broken down by saliva into what are called tastants.[68]

Taste receptors then convert information from the tastants into electrical signals that are carried to the part of the brain where those impulses become taste. Collectively, this is known as the gustatory cortex. "It is important to know how the brain creates the flavors that we perceive," writes Napa Valley College professor George Vierra. "Foods do not contain flavors. They contain flavor molecules. The flavors of those molecules are created in our brains."[69]

In addition to taste, the tongue is also the primary sense organ that registers feel (such as texture, density, chewiness and oiliness), temperature and spiciness. Spiciness is not a taste—but a pain signal—and isn't detected by taste receptors.[70]

The tongue is both a muscle and a gateway, one of our most ancient forms of protection. On my first day tasting cacao (when I couldn't be sated), I said yes to pod after pod. Sucking off the sticky sweetness and spitting out the bitter seed was evolution at work. My tongue and mouth were doing exactly what they were programmed to do—and my capacity to appreciate and understand biodiversity grew.

While Brad oversaw the installation of the flavor lab, I joined Brad's colleague Katie and Javier junior and senior on a trip to the grocery store and a few smaller markets in Esmeraldas. They wanted to buy local foods so that when Brad and Katie were sitting back at TCHO's headquarters in California, they could talk to farmers at the co-op about something like cacao's astringency in ways that would resonate with everyone. Brad explained, "A big part of what we're trying to do with the flavor lab is develop a communal language of how the cocoa is perceived—and what creates value."

We stocked up on items that could help build those connections across cultures: dried and fresh fruits, nuts, flowers, cigarettes, vinegar and aspirin. "Flavor is a language," Katie reiterated as she dropped oranges and lemons into her basket. "And every place has both its own language and dialect." This reminded me of the work of Linda Bartoshuk, an acclaimed taste researcher and director of psychophysical research at the University of Florida's Center for Smell and Taste. "When it comes to food tastes," she says, "we all speak in different tongues. People inhabit separate taste worlds."[71] There isn't one definition of value or deliciousness, but, as Ann Noble taught us, a common language can be transformative in helping us interpret flavor.

When we got back to the UOPROCAE compound, Katie started to lay out items according to taste and intensity, from acidity and astringency to bitterness and earthiness. We sorted tomatoes, pineapples, lemons, vinegar and *naranjillas* (tart, tomato-like fruits) to explore acidity. In another pile, we clustered peanut skins and underripe bananas for astringency. Aspirin diluted in water at varying strengths was how we tested for bitterness, while loose tobacco, rocks collected from the street and dirt from the side of the road served as our benchmark for *terroso* (earthiness). The defining aromas of Ecuadorian cacao—floral and nutty—were represented by fresh roses and jasmine flowers stuffed into plastic water bottles, plus an array of peanuts, hazelnuts and almonds.

Our first goal was to affirm that smells and tastes vary in intensity. The second was to understand that, once we started to pay attention, we'd be able to distinguish between features we'd likely never considered. "Is the peanut skin *amargo* [bitter] or *astringente* [astringent]?" Katie asked. Bitter is a taste; astringency is a sensation. "Is it a sharp unpleasant flavor [that's bitterness], or does it make your mouth and tongue feel dry [that's astringency]? Notice how the acidic fruits make your mouth water," she said. "Ask 'What is happening?'"

The following afternoon, we tasted liquor (the thick paste of ground cocoa nibs) made with cacao grown by UOPROCAE farmers and processed in their new flavor innovation lab. Tasting liquor is an intense experience—the equivalent of a shot of espresso (sans sugar) versus, say, a cappuccino with sweetener. It's unadulterated cocoa—nothing more or less—the easiest way to understand differences in flavors. Katie cautioned against describing the cacao as good or bad but, again, asked, "What is happening?" reinforcing the strongest skill to hone was observation.

She then brought out additional liquors from Peru and Ghana, as well as ones with very clear defects that tasted metallic, moldy and dusty. Even though she had cautioned us against judging the earlier liquors, here she was clear: These were bad. "In order to know what's good," she said, "you also have to know what is not—and why."

Cacao from Papua New Guinea, for example, is often dried on beds heated by diesel or wood, both of which can impart hammy or oily tastes to the beans because fats absorb odor. These are defects, the kinds of things most chocolate makers, particularly those concerned about flavor, *don't* want because they mask aromas inherent in the beans.

Yet one maker (who has gotten a lot of media attention and puts its chocolate in the most beautiful of wrappers) has decided to turn this defect into an attribute, repackaging the off-flavor as a novelty by highlighting the smokiness of the bar. Many craft makers who work closely with farmers on improving drying techniques and eradicating those off-flavors question if this is something we should celebrate—if, by buying into the smoke, it's making it harder for producers who are trying to improve the taste of their beans. This is a question only we, the eaters, can answer, but it's important to recognize we're vulnerable to external influences, including hype and packaging.

A recent study by Hilke Plassmann and Bernd Weber reaffirmed our vulnerability: We tend to enjoy identical products more when they're priced higher or highlight positive "expectations of ... pleasantness."[72] This doesn't just happen in our mouths and noses but also in our brains.[73] We are influenced by what the world tells us to like. People who rave about the flavors of raw (fermented but not roasted) chocolate likely don't know that some of the finest attributes of the bean are brought out during the chemical reactions that occur through roasting—the step raw chocolate makers forgo.

This isn't to say we shouldn't try a wide range of chocolates, but if we're going to spend time and money and take in calories, we should know what our investment is supporting. We should try to understand where the flavors come from—and what good farming and processing practices taste like in order to tease out why we love what we love. To find out what's happening.

Ecuadorian chocolate is, and will always be, my first love. This isn't because it was the first chocolate I ever tasted, but because it was the first one I tasted deeply, with full attention. After three weeks of stuffing myself full of *baba* and sampling liquors from all over Ecuador, I finally found the floral taste that had eluded me.

I met Maryuxi Espinoza in the final 24 hours of my three-week trip. Maryuxi—chief of product development and sensory analysis for Transmar, one of the largest cocoa exporters in Ecuador—samples every single batch of cocoa that leaves the facility, including the cocoa destined for TCHO. She knows Ecuadorian cacao.

The next to last sample of liquor Maryuxi shared with me had a flavor, I later learned, she held dear. I expected toasted nuts and a light bouquet, but I got a burst of violets: green stems, fragrant flowers. "Es floral," she said. Yes. This was floral. This was Ecuador.

Ecuadorian cacao is among the finest in the world. The country is the birthplace of what's known as Nacional or Arriba cacao. *Arriba* (meaning "up") is a reference to cacao that comes from the upper basin of the Guayas River, as well as all the rivers and streams that feed into it. The genetic grouping also includes other diverse varieties from the region.[74] It's this tapestry of flavors that makes the country one of the top global producers of specialty cacao.

"Chocolate is a part of us; it is a part of Ecuador," explained Maria De Lourdes Alvear, an analyst at Agrocalidad, the government agency that regulates and oversees agriculture. "Yes, we are a small country, but we have something no other country has: the flavor."

In wine, variations are expected. Scott Schultz's Trousseau Gris reached for me and made me pay attention, but the notion of terroir and diversity weren't entirely foreign concepts. In chocolate, they are—or, rather, they were. I didn't expect this experience, not with the pulp or roasted beans or with the end bar (what Darin Sukha once described as "the continuation of a journey that has already started").

The liquor Maryuxi shared with me was a surprise. While the bouquet I would later find in my coffee from Ethiopia was made up of fragrant white flowers, here I found earthy purple ones: a bed of wild violets, stems and all. I was stunned.

The beans were grown by Mina Bustamante de Caicedo, an 81-year-old woman from Vinces, a city known as "Little Paris" that boasts a mini replica of the Eiffel Tower and a rich history of cacao cultivation. This diversity found in Mina's cacao is what we're losing, in large part because of the diseases that plague cacao, but also because of the high-yielding variety that has been bred to resist them. You see, the country that has given the world the majority of fine flavor cacao is also the birthplace of CCN-51, a variety that is slowly but surely wiping out diverse flavors—not just in Ecuador but throughout the world.[75]

But, before we delve into CCN-51, we have to understand the reason it exists. That takes us right back to the steamy environments where not only cacao but diseases thrive. Black pod rot, witches' broom, frosty pod rot—they are some of the sinister-sounding fungi that, along with mirids, moths and other insects, destroy up to 40 percent of the world's cacao crop annually, with losses estimated at $2 to $3 billion.[76] "Whether in a human or a plant," explains Surendra Surujdeo-Maharaj, plant pathologist at Trinidad's Cocoa Research Centre, "in order for disease conditions to exist and thrive you need a susceptible host, an aggressive strain of pathogen, and the right environmental conditions. Cacao farms have all three."

Black pod rot, which results in "more production losses globally than any other disease of cocoa,"[77] is caused by several species of fungi within the *Phytophthora* genus, a reference to the Greek term for "plant destruction."[78] It starts in the soil, goes up through the roots and, Surendra adds, "is capable of infecting all organs of the tree." It's awful: The mummified pods don't fall to the ground and disintegrate; they hang from their tiny stems, shriveled and black.

Witches' broom—the most ruthless cacao disease in Latin America—looks a lot like it sounds: tangles of misshapen shoots that seem to suffocate the tree and can lead to its death. In 1989, witches' broom was introduced to cacao-producing areas of Brazil in what was widely believed to be a deliberate act of biological terrorism.[79] Less than 10 years later, the country went from being the second largest cocoa producer in the world to an importer of beans.[80]

Witches' broom is unyielding; the tangles of shoots are gnarled and tough. The only way to get rid of the brushlike clusters is to cut each one out by hand. Otherwise, the disease will "turn into a spore grenade," explained Gualberto Valdez, a coordinator at CEFODI, an NGO dedicated to rural development. He told me this as we walked through cacao farms in Esmeraldas, gesturing broadly. The fungus will spread "en todas partes," he said. *Everywhere.*

While witches' broom is more prolific, frosty pod rot is even more devastating. *Moniliophthora roreri,* the fungus responsible for frosty pod,[81] starts off as a lesion and completely devours the pod. There is little farmers can do; the only remedy that seems to work—copper fungicide—is too expensive for most of them.[82]

Diseased pods look like they are dusted with powdered sugar. It's evident that the lightest of touch could send those fungal spores flying, carried wherever the wind might take them. First identified in Ecuador back in 1918,[83] by the 1920s, frosty pod rot, along with witches' broom, had decimated nearly 70 percent of Ecuador's cacao.

Farmers were desperate for a solution. They found one, decades later, in the hybrid CCN-51. It was developed by agronomist Homero Castro, a man who had spent years crossbreeding different cacao plants he'd collected from his travels. In 1965, on his 51st attempt, Homero crossbred plant material from the eastern jungles of Ecuador with one clone that originated in Trinidad (ICS-95) and another that had originally been collected in the Peruvian Amazon (IMC-67).[84] It proved to be a winning combination. He named the variety CCN after the farm in Naranjal where he had conducted his work—Colección Castro Naranjal—and 51 for the number of times it took him to get it right.

"Homero Castro was a scientific rogue," cocoa and chocolate researcher Cristian Melo asserted when we met at a café in Quito. Castro left the national institute of agricultural research (INIAP, Instituto Nacional de Investigaciones Agropecuarias) to start his own research program. Working alone on his Naranjal farm—aptly called Theobroma, after cacao's botanical designation—he developed a high-yielding, frosty-pod- and witches'-broom-resistant variety. But, sadly, he never experienced the full glory of his work. Castro and his daughter were killed in a car accident in 1988 when leaving Theobroma. As a result, CCN-51 was never patented, and ended up being released into the wild and planted on farms without much oversight. "In less than 30 years," Cristian explained, "CCN-51 became the pride of the Ecuadorian cocoa industry."

How did this shift occur? I asked. And why? Cristian clarified: "In terms of yield, CCN-51 was like a sports car compared to a minivan. It produced four times as much as Nacional but the price for both varieties was the same." The promise of yields, disease-resistance, plus a lack of oversight over what was planted, resulted in the hybrid's rapid spread. Farmers systematically replaced traditional varieties with CCN-51.

"They took out everything," Francisco from UOPROCAE told me. "Anything old was destroyed in favor of these big pods." Cristian added, "If you wanted Nacional during those early years, you had to get it from INIAP. Outside of the research institution, there was no place to get traditional varieties."

The very first CCN-51 acreage Homero planted was on a farm owned by the grandfather of Sergio Cedeño, the manager of Hacienda Cañas. "My grandfather was skeptical," Sergio told me. "So he only let Homero plant 5 hectares [roughly 12 acres]."

I met Sergio shortly after landing in Guayaquil, the heart of Ecuador's cacao industry and the port city from which all of the country's cocoa is shipped. Sergio offered to pick me up from the airport and take me directly to his farm. I thanked him, sent him a photo so he'd know whom to look for, and asked what he looked like. He told me he'd be the one in the Montecristi hat (commonly known as a Panama hat, though actually developed in Ecuador).

These cultural distinctions are important to Sergio. He takes tremendous pride in not only his history with cacao ("I am the fifth generation in my family to grow cacao, starting with my great-great-grandfather in 1860"), but also in the agricultural history of Ecuador. It is why he dedicates a portion of his time to cataloging and saving tropical fruits native to Ecuador in a small *ex situ* collection.

Yet, Sergio also manages the largest exclusive plantation of CCN-51 in the country, the breed that has rapidly and systematically eroded the diversity of Nacional. While I find Sergio's native fruit collection to be at odds with this larger picture of genetic erosion, he does not. Sergio is trying to save what's special about Ecuador, he says, including the farmers and laborers who grow and harvest cacao.

Within minutes of stepping onto Sergio's farm, I was presented with a full glass of juice from the pulp of CCN-51. Its abundance was immediately evident: CCN-51 fruits are big and juicy, with large seeds that hold a higher percentage of fat, which, once processed, become lucrative cocoa butter.

Sergio led me through rows and rows of small trees dripping with purple pods. "CCN was bred to be short," Sergio explained, "so it's easy to prune the tree and easy to harvest." I thought back to jabbing the *palanca* toward the 10-foot trees on Alberto's farm. These pods would be a lot easier to cut down once they ripened (to a reddish orange). And any diseases would be a lot easier to manage.

But this workhorse variety isn't without its drawbacks. The challenges of monoculture cultivation are the same with every crop: They increase risk. One pest, one disease, can wipe out everything. In the wake of cacao losses, bananas replaced cacao as the leading export crop of Ecuador. By 1950, the country was the top producer of bananas in the world.[85] But, again, disease hit: Panama disease (Fusarium wilt) wiped out the Gros Michel banana that was most widely grown. Those fields were eventually replaced with the Cavendish banana, but now that's succumbing to the Tropical Race 4 strain of *Fusarium oxysporum*.[86] As Bernie Prins and Peter Fanucchi explained when we walked through the genebank's vineyards, this is the risk farmers have to take. Industrialized agriculture demands farmers grow as much as possible as quickly as possible and deal with devastating consequences only when they occur.

Unlike traditional varieties, which need shade early in the growing process,[87] CCN-51 was bred to grow in full sunlight and requires more water. It also needs more fertilizer, which is typical of high-yielding varieties. And, as researcher Ed Seguine told me, "it's a crop that really only works as a monoculture because it has to be dried and processed separately."

Despite this awareness, that's not happening to the extent it should. Most batches of CCN-51 and Nacional are not being separated as required. This has resulted in cacao that's compromising the diversity found in Ecuador, along with the overall deliciousness of the crop.

"The funny thing is that nobody really cared about quality," Cristian explained over steaming mugs of drinking chocolate. "Not the farmers, not the exporters. CCN-51 needs more time to get proper fermentation—seven days instead of three. When farmers mixed CCN with Nacional, they got the worst of the two: Nacional was well-fermented, but CCN was not. Also, CCN-51, when properly fermented, has a good basic chocolate taste—but that's it." He tipped the last sip of chocolate into his mouth and added, "In taste terms, CCN is a cheap perfume, while Nacional is Chanel No. 5."

During my tasting tour with Maryuxi, I asked to sample liquor made from CCN-51. I had tried both *baba* (in the form of juice) and liquor on Sergio's farm. The juice from the pulp was tart and sweet; the liquor tasted like cocoa. I liked them just fine. But I wanted to try again, away from Sergio's watchful eye, in a more neutral environment. Maryuxi gave me three different samples and, again, I didn't think they were bad. The liquors were a little metallic and acidic, but still tasted like chocolate, nothing more or less. Maryuxi concurred. "It's not bad, but it's not interesting.

It's flat. It can be used as a base, and, with sugar, it will make chocolate. You don't need something special for a Hershey's bar."

To be fair, CCN-51 was never intended to have the floral notes or flavor complexity found in Nacional. Homero Castro selectively bred it for disease resistance and yield. He wasn't thinking about taste because, at the time, no one was. But now farmers, manufacturers—and even conservationists—are.

CCN-51 grows well in many places. It isn't interesting or special; it's ubiquitous and ordinary. As Alberto said, this leaves Ecuador competing with every other country growing bulk cacao. The difference is that Ecuador has the capacity to offer something special—something places like Ghana or Ivory Coast will never have. And that specialness is being lost, pod by pod.

"We don't only want CCN-51," Sergio stressed. "We want a diversity of cocoa with *high* production. With low production, people live in misery, in poverty." When he said this to me, we were driving down the long street that led to Cañas farm. He pointed out the sturdy concrete homes of the laborers who worked on his land contrasted against rickety bamboo structures. "This is what CCN does. You cannot progress with the old Nacional varieties."

This is because prices don't reflect the real value of the crop. We want chocolate—and we want it to be affordable. Farmers want to satisfy our need, but they also have to be able to earn a living. "So the tension remains," Cristian says. "On one side, you have a high-yield variety that gives you lots of cocoa with good chocolate flavor. On the other, you have a low-yield traditional crop: one that gives a cocoa with an *incredible* taste, the kind of thing craft chocolate makers dream about. The problem is, most times, farmers get the same price for both. In the end, we are asking Nacional farmers to forfeit the profits they would have made if they had switched varieties."

Although INIAP has now developed 10 more clones that offer improved yield and taste, Cristian says, "They have come years too late. Farmers know CCN-51 is a productive variety; they trust that. So now they have to also overcome the fear of trying something new. The rewards take years to manifest, and farmers wonder if they should risk their money on something that may or may not work."

To support these farmers, we consumers also have to try something new. That means reaching for a specialty or single-origin bar instead of a candy bar, when possible. Otherwise, CCN-51 will continue to erode the genetic diversity of cacao in Ecuador and the world.

This genetic erosion has also impacted Ecuador's global status as a cocoa-producing country. In 2005, the International Cocoa Organization (ICCO) downgraded Ecuador's status from a country offering 100 percent fine-flavor cacao to 75 percent.[88] "And that was generous," explains ICCO executive director Jean-Marc Anga, who believes the amount of CCN-51 grown in Ecuador exceeds 25 percent. "We have to protect the purity and integrity of fine flavor."

Protecting that integrity is more challenging than it seems, primarily because we don't pay for its value. Think back to the moment when Aldo reached his hand into the bag of fresh cacao at UOPROCAE and found bark, rags and rinds that were used to increase its weight. There's little financial incentive to improve flavor—or grow more Nacional.

Because we haven't yet figured out how to successfully ferment small batches of cacao, a lot of it is sold to intermediaries. These middlemen collect small amounts from individual farmers and then process it all together. "Intermediaries have a lot of power," Transmar's commercial supervisor Nadia Rosales says, "because as soon as the cacao pod is harvested, it starts fermenting. You have to sell the crop."

Nadia and I first met in late spring of 2014 when she took me to Transmar's collection center in Naranjal (which accepts beans after fermentation) and to nearby farms. When I asked if we could also meet some of the people behind the "Compramos Cacao En Grano" ("We Buy Cocoa Beans") signs, she was nervous. Intermediaries operate in shadier spaces than official commercial exporters. Some offer high-interest loans to farmers and wield that financial power like a blunt instrument. They also have a low threshold for quality, which is how they get away with purchasing cacao for less than the going rate.

When Nadia and I paid them a visit, both Nacional and CCN-51 were selling for $132 per quintal (100 pounds). "There's a $1 to $2 difference in high season," she said, "nothing more with the middlemen."

I asked one man, with thick gold chains around his neck and an extremely voluptuous woman on his arm, if he separated the cacao. He insisted he did. But when he stepped away to take a call, Nadia whispered for me to look at the size of the beans: Nacional and CCN-51 were being fermented together.

When you ferment super-pulpy cacao beans with ones that have a lot less pulp, you end up with a cacao that tastes terrible. This is a lot more common than we realize because once manufacturers add in sugar and fat, along with pretzels, caramel, peanuts and other ingredients, the true taste of the cocoa gets lost. It disappears in the chocolate and, eventually, disappears from the place where it was grown.

So is it enough for chocolate to be chocolate-like? Do we have to settle? A recent piece on National Public Radio said yes. They called CCN-51 the hybrid plant that would "save chocolate," adding that, if we wanted chocolate at all, we'd just have to get used to chocolate that tastes like "rusty nails."[89] Although this is far from our only solution, it is the truth we're heading toward.

How can we make the case for farmers growing anything other than what is most abundant if there is no compelling financial reason for them to do so? Are they supposed to keep doing it for honor—for the integrity of the cacao?

"The number one complaint we get is on flavor," Nadia admitted on our drive back to Guayaquil, "but no one wants to pay for it. This is the cacao of Ecuador, of my people. This loss hurts me, too." Cristian pushed further: "How much money are people willing to pay for flavor? Twice as much? Three times? And how much of that goes back to the farmer?"

These questions don't have simple answers.

Should we sacrifice flavor? Is it time to give up and settle? According to the farm and the farmer I visited immediately after Hacienda Cañas, the answer is no.

The midges and mosquitoes were, again, relentless: buzzing in my ears, biting through my jeans and shirt, leaving angry, itchy welts. But I was undeterred. I doused myself with DEET and proceeded, following the farmer deep into the forest, greedily saying yes to pod after pod, stuffing the fresh cacao into my mouth until my cheeks bulged, spitting the seeds into my increasingly sticky hands.

But instead of dropping the seeds to the ground when they started to spill out of my hands, I carefully placed the seeds—the future chocolate—into the empty half of the pod. I had to. It was required. Everything on the farm was meticulous and organized—just like its owner, Vicente Norero.

I met Vicente, the proprietor and manager of Camino Verde, during Semana Santa, a week of penance preceding Easter. In Ecuador and other Latin American countries, this repentance is visceral. It involves not only sacrifice of something you love ("I'm giving up chocolate for Lent"), but a physical manifestation of suffering. Men walk barefoot through the streets with crowns of barbed wire on their heads, approximating the crown of thorns Christ wore, and drag heavy wooden crosses, symbolizing the Crucifixion. They go deep into the pain of loss before readying themselves for the joy of rebirth.

Thirty-six hours before watching these men, Vicente showed up—unannounced—poking his head into my hotel room as the receptionist and I tried to get the air-conditioning to work. I had $5 plastic reading glasses perched on the edge of my nose and was waving a remote in the direction of an AC unit. Then Vicente peeked in: tall and blond, in a crisp, light blue shirt and slim-fitting Euro trousers.

"May I help you?" I asked, pushing my glasses into place.

"I'm Vicente Norero." My vision of a cacao farmer was more like that of Javier senior or Alberto, not an Italian-Ecuadorian man in designer pants.

We confirmed plans to meet the next day. He had originally asked me to meet him at his office, but once it became clear I had a terrible command of Spanish and an even worse sense of direction, Vicente suggested we meet on the street corner, just down from the hotel.

The next morning, he called the reception area where I was eating breakfast and told me, "I will see you in 15 minutes." I asked him for 20. I quickly finished my eggs, packed my bags, checked out, went down to the corner and waited. Twenty minutes stretched to 45. I had no local cell phone or way of contacting him from the street. I weighed the idea of dragging my suitcase back up three flights of stairs to the hotel to use their phone, but worried I might miss him. So I waited. And waited.

Just when I was on the verge of giving up—a theme of nearly every visit with Vicente and, maybe, of my life—he showed up. He apologized profusely, explaining the snarl of traffic around the parade of people honoring the Crucifixion had slowed him down. I was visibly frustrated. He knew I had wanted to see the procession.

So he drove back into the snarl to try to assuage me. It worked. "Go there, around the corner, and you'll see the people. I'll wait here," he said. I stepped out and slowly walked away from the car, my fear of losing Vicente overridden by my desire to experience the dramatic weight of the moment. As I watched the barefoot men and women, I realized there was no better time than this to meet Vicente.

We all carry the weight of our suffering, some more visibly than others. Only one culinary companion has lifted me out of those deep wells of sadness: chocolate, the substance of my wedding cake and my divorce. For many, it is considered an aphrodisiac—the stuff of Valentine's Day. I also consider it love, but not the kind that comes in a heart-shaped box. Chocolate is nothing like the coffee I have in a café or the wine I share with friends. There's no "we" in chocolate; this steady companion is mine alone.

I walked back toward the car. Vicente was right where he said he would be. "So," I asked, "how is chocolate your redemption?" It seemed like a strange question, but Vicente got it. His relationship to cacao and chocolate also goes far beyond the substance.

Vicente is 44 years old. Chocolate has been his family's business since 1889, when the Norero-Ceruttis founded La Universal, which became one of the most lucrative chocolate factories in Ecuador. But his parents always encouraged Vicente to do something else. "I wanted to be a pediatrician," he told me. "And my father wanted me to be an engineer. I ended up studying economics instead, and wrote my thesis on the economics of cacao."

The company rose to prominence in 1906 but, due to a number of factors (including Ecuador's economic crisis), shut down in 2002. "When the company closed, it was like having *lepra* [leprosy]. Nobody wanted to talk to us. They'd cross over to the other side of the street when they saw me coming." So Vicente began again—from the bottom up. "I worked with a company in Quito and learned, little by little, about exporting. It was hard, but I knew I had to come up by my own resources. I started investigating, studying—and remaking myself."

Through marriage, Vicente ended up taking over an old cacao farm. "The cacao plantation was in the same situation I was: fucked up. But I fought for it, and together, against the odds, we transformed." Over years he expanded the farm into what is now Camino Verde, named after the long, verdant path that runs from one part of the farm to another. "One of the most frustrating things I hear again and again from farmers and chocolate makers is that this work 'is easy when you have a farm like Camino Verde.' What they don't realize is that I started from an abandoned 16-year-old farm; they don't know what I have done for this. Fighting against the odds, sacrificing and working hard—that has been my redemption."

Vicente oversees multiple farming operations, growing bananas, corn and both CCN-51 and Nacional. He is trying to balance the need for scale and quality: His Nacional cacao is sold to Rogue Chocolatier and other craft chocolate makers, while his CCN-51 is exported for cocoa butter.

"When I was a kid, I never liked the confectionary shit. I liked the real chocolate. I have memories of these rich, dark chocolate Easter eggs I used to eat, and I want my kids to have that same experience." While he acknowledges he can't maintain the life he wants growing only Nacional, he refuses to let it go. "You have to sacrifice something, but it will not be quality. It will not be the respect for the cocoa beans."

What Vicente is trying to do is find some middle ground, and make his traditional cacao more interesting by pulling out certain flavors. He grows plants he thinks have enhanced nutty characteristics all together and separates out anything that leads with floral notes into a different area. (That's why I couldn't allow any seeds to drop to the ground.) He also introduces ingredients and enzymes to his fermentation process to layer on additional flavors: chilies to make the cacao more *picante*, banana skins to impart a banana aroma.

Although the incredible mosaic of flavors inherent in Nacional are muted because of these processes, for Vicente, it adds something special. "I am not trying to do Nacional," he says. "I am trying to do Camino Verde." His ambition is raw and undeniable; he is trying to find a way to make his cacao unique and get the price premiums that will allow him to keep going. To Vicente, this means ensuring every batch of cacao will have some consistency—helping craft makers, like Colin Gasko of Rogue, manage the tension between the dynamic nature of cacao and the standard required for chocolate.

"You are eating a story," Vicente says. "Mine is the story of Balao, which is, ironically, also the place where frosty pod rot started in Ecuador. You see, both the farm and I have good and bad that we're trying to redeem. In the Balao bar from Rogue and others, you are eating *my* story: my history, my work, my hopes, my legacy."

And in TCHO's chocolate, we are eating the fruits of Alberto and Alexandra's story. As I listened to Vicente, I recalled watching their 3-year-old daughter, Anai, stuff pulpy seeds into her mouth with the same fervor I did, and hearing Alexandra say, "We live off this. We fill it with love and care because it's coming from our family."

We are also experiencing the stories of Brad and Colin. Every bar holds narratives, but, in most, the stories become muddled: farmers remain unnamed, workers remain unseen. The cultural and genetic diversity disappear.

The next day, Vicente and I met—on time—and headed back to his farm in Balao. We walked the expanse of land where his organic cacao is grown, the only part of the farm that felt fully wild. But as we climbed over rotting logs and weaved between giant spider webs and overgrown plants, I saw Vicente still had a plan; like Alberto's farm, this chaos was organized. "I am reading the trees," he said.

He later wrote about our experience: "Each pod has its own story— of how the sun and rain touched them, of times they were sick, of how they became healthy. And their journey continues as they go through fermentation and drying to become the perfect cocoa bean."[90] I realized this is how one bean becomes more precious than another. It is treated with reverence; it is supported in living up to its potential and displaying its best qualities.

But to even consider this is a luxury. Not every farmer cherishes this work. Carla Martin, a lecturer at Harvard University who teaches classes on chocolate and food politics, reminded me, "There is a sense of disaffection when people grow a crop they don't eat. When you work on a Cadbury farm in Ghana making 30 cents a day, why care if the cacao is fermented well or dried properly? It's hard to connect with a crop that's so abusive to you."

Vicente is not your average farmer. First, because he is meticulous to the point of obsession: He takes no shortcuts; he never thinks of cacao as just another crop. Second, because he lives in a mansion bigger than most I have seen on any continent and travels to his farm by plane (it's faster, he says; he can get to work sooner). Vicente has a large enough farm that he can ferment his own cacao, and he has access to the chocolate makers who can help shift the tide. But his wealth can't insulate him from floods or frosty pod rot.

He, too, has a lot to lose.

As we drove back to my hotel in Guayaquil, the radio blared Toni Braxton's "Unbreak My Heart" in Spanish, and I thought about the hunger we have for both love and chocolate. How some of us do settle: for security, for status, for the safety of knowing something or someone is there. My question shifted from "Do we have to settle?" to "What does settling mean?" That answer is personal to each of us, but when something exquisite is so close, shouldn't we wait for it and fight for it? For me, the answer is clear: Yes. If I had wanted to settle—in life, in love, in chocolate—I would have done so a long time ago.

The same holds true for Vicente. As we pulled up to the curb and the hotel doorman reached to open the car door, Vicente turned to me. "I am just a guy—one of many farmers who fight every day for cacao beans. But I don't want to be anonymous. It's easy to arrive, but it's difficult to maintain your position; I haven't even arrived yet." Later, Vicente emailed me to underscore his point. "This is something I deeply believe in," he wrote. "I can't accept the idea that, because this is hard, we have to surrender to bad chocolate and bad food in general. Cacao is passion, it's history, it's mystical. When we standardize cacao, it's the same as killing its magic, killing the opportunity that kids and future generations will have to feel all that happiness—to taste all the flavors we grew up with when we tasted real chocolate."

That night, while I stood in the hotel bathroom washing dried cacao pulp off my jeans, I thought about what renowned chocolate expert Chloé Doutre-Roussel said when I asked her how we could save chocolate: "The biggest level of change we need in cacao is the freedom of choice." The farmer needs to have enough knowledge and options to do what he needs and wants to do, to "choose consciously."

For Vicente, his cacao farm is about redemption and responsibility. Redemption for the chocolate factory his family lost and responsibility for future chocolate lovers: "This is something that a kid will put in her mouth. There is no flavor without work, not just in cocoa but in any food. There is no chance of something great without sacrifice."

For me, the lessons of cacao and chocolate are those of reclamation: of embracing what is wild in me, of allowing for a deeper yearning, of cherishing myself in times of lower productivity—and remembering that honoring diversity, in ways that are both real and symbolic, allows for the potential of something greater.

A bite of something sweet can show the world we don't have to settle. We can have something exquisite.

III. TASTING CHOCOLATE

Everything is a sensory experience. Have confidence in
yourself. You've been doing this since you were born.

—DARIN SUKHA

*Free yourself of distractions, and try to observe yourself as you walk
over to the chocolate aisle.* The experience of tasting starts long before
the chocolate reaches your hands. What emotions do you bring to the
moment? What does it feel like to take a chocolate bar off the shelf and
hold it in your hands? What drew you to one chocolate over another? The
weight of the bar or the texture of the wrapper? Pay attention, but also
understand the cover reveals just one part of the story. Chloé Doutre-
Roussel argues nothing should be on the label because we tend to be
influenced by information that doesn't shed light on flavor: "The percentage or place, in and of itself, does not tell you if the chocolate will be
delicious or express the full potential or terroir of the bean."

For a tasting, select four or five bars from similar origins or makers.
Chloé says, when you explore the world of chocolate, you fall in love
with a maker's style (one that also reflects the values and practices of the
company): "Start with one and buy more bars from that brand."

If your selection is based on geography, include a chocolate from
Ghana, the global baseline for chocolate tasters and researchers.
Research origins so you know what to expect, or explore a few different

makers' interpretations of a single country. "And ask for advice from a specialty shop run by people with passion and experience who select their products and know all about the brands and the bars. Don't look for advice online," she cautions.

There is no right or wrong way to taste, but as with wine, try to taste in order of intensity so you don't overwhelm your nose and mouth at the start. A general rule of thumb is to go from dark to milk chocolate, and from highest percentage of cocoa mass to lowest. But the most important thing is to take your time: "Because of the slow melting properties of cocoa butter and the unparalleled complexity of cocoa, chocolate flavors can linger for 20 minutes or longer," Brad Kintzer explains.

Milk chocolate has a lower cocoa content and includes dairy, which, Brad notes, can add in "a whole world of flavor because you're caramelizing lactose." There aren't that many milk chocolates on the market that are made to be explored deeply, he says, "so try the dark." If you opt for a higher concentration of cocoa mass—85, 90 or 100 percent— you're reducing the sweetness that comes from sugar and increasing bitterness. "And bitterness is only one sensation. Open your mind—and get ready for a flavor fire hose to the mouth."

Get a glass of warm water and water crackers (made with just flour and water, no salt) to cleanse your palate in between tastings. When you chew the crackers, nibble them with your front teeth; don't let them get stuck in your molars. Swish the water around your mouth as needed, and allow yourself time to reset between different samples of chocolate. The slower you go, the more you will experience.

Let the chocolate warm up. This shouldn't take too long because it should already be at room temperature. You never want to refrigerate good chocolate unless it's the dead of summer (or you live in the tropics), as it causes the cocoa butter to separate from the solids and creates the whitish discoloration known as bloom. This doesn't impact flavor, but it isn't pretty.

Peel back the paper and open the foil. There's the chocolate. What do you see? What is the story written on the chocolate? Is it glossy or dull? Thick or thin? Does it have a design (Dick Taylor chocolate bars look like filigree) or a logo (Patric bars feature his signature)? Are the squares already apportioned for you? Are they big, like Marou, or small, like Bonnat? Rogue Chocolatier bars have no squares—you get to decide how big or small of a piece you want. Keep looking. Does the bar have any evidence of bloom or any air bubbles? Although they won't affect taste, they reveal the kind of attention that went into making the bar.

The wheel found on page 345 was created by Seneca Klassen. His reason for putting it together resonates with Ann Noble's impetus for the wine wheel: "Whenever you're building a deeper understanding of any new food or beverage, it's helpful to have a linguistic beginning point for that exploration. The wheel isn't meant to be exhaustive; it's a tool to provide a starting place."

Hold the bar up to your ear and snap off a piece. Listen to how it breaks. Set a small piece into the crease between your index finger and thumb. Wait patiently for it to melt slightly. See how it feels. Through the warmth of your touch, the cocoa butter that was solid at room temperature is starting to melt.

Look at the chocolate. It's not one color: A chocolate from Ghana has deep brown tones, while one from Madagascar will be more of a reddish-gold color. If the chocolate has milk in it, it will be even lighter.

Keep the chocolate there in the crease of your hand and bring it toward your nose. Notice how the aroma increases as it gets closer. Create a small smell chamber by cupping your other hand lightly over the chocolate. Pay attention to what thoughts arise, what memories surface. Allow for all of it. When Brad guided me through a tasting, he kept encouraging me to have what Buddhists call "beginner's mind." "Don't even think of it as chocolate," he said.

Inhale. You know from our wine exploration that flavor is a combination of what we perceive in the retronasal cavity as well as on the tongue. What's there? If you have Buddhist restraint, inhale again (quickly, in sniffs). Reference the Chocolate Flavor Wheel. What aroma is dominant? What is secondary? Has anything new surfaced? Is the aroma strong or weak? Enduring or fleeting? Write down what you're experiencing.

Now—finally—put the chocolate in your mouth. Your tongue is an instrument of taste. Close your eyes. Let the magic happen. "Tasting chocolate is a completely unique sensory experience," Brad explains.

"Through genetic selection, fermentation, drying and roasting, we convert the seed that was designed *not* to be eaten into one of our most dense, flavor-packed foods."

Let the chocolate melt on your tongue. Notice if the melt happens quickly or if it takes time. The thickness of the bar impacts the speed and sensuality of the experience. Observe the coating that remains in your mouth: Is it creamy, greasy or waxy? "Wine and coffee have their own viscosity and mouthfeel," Brad reminds us, "but they are nothing like chocolate; they don't have the slow delivery system."

Because the aroma molecules are volatized and travel by air, inhale through your mouth as you coat your tongue with the melted chocolate. Gauge its texture: Is it silky or grainy? Assess the finish: Does it linger or end quickly?

If you wish, spit the samples out instead of swallowing them. (We are, after all, tasting, not eating.)

Break off another piece. Embrace beginner's mind and taste anew. What impacts you first? What is the intensity of the initial experience? And the next, and the next, and the next? Ask this question continuously as the flavors reveal themselves to you.

And don't get discouraged. Chloé is credited with having one of the most sophisticated palates for chocolate in the world, but she is adamant: "I am not an expert either. I am someone who has learned by doing. The only difference is that I am 47; I've tasted a lot, and I have a good sensorial memory."

Let the tasting be the beginning of a bigger sensory experience. "Write it down, talk to people about it and have fun with it—and then do that with everything. Do it with your cereal in the morning, do it with your juice: cross-train. Tasting will become a habit," Brad says. "So much of our sensory training and appreciation isn't just about chocolate but other foods as well. It's about paying attention to what we're doing. It's about looking deeper into the everyday."

Taste it all.

COFFEE

I. THE CUP

Within ten minutes of waking up, I'm making coffee. Yawning, stretching, wiping sleep from my eyes, I shuffle toward the kitchen and put the water on to boil. I pull the milk out of the refrigerator and, out of the cabinet, a container holding small brown beans. The beans are matte, most revealing a thin ivory split, evidence they were washed before being dried and roasted. I pour them into my hand and inhale, then funnel them into a palm-size grinder. Filling the metal cylinder with beans, I press down the cap, fasten the removable handle to the bolt on top, and turn and turn and turn.

I am making coffee.

Hand grinding takes all of three minutes, but every morning, as I rotate the handle that turns the burr grinders that break up the beans, I grow impatient and consider replacing the manual grinder with an electric mill. And every morning—immediately following that wish for more ease and speed—I send out a cosmic thank-you for the farmers who planted, nurtured and harvested my coffee. The inner proclamation is quick but sincere. The work I'm doing, in this moment, is nothing compared to theirs.

Guides on making the perfect cup of coffee never tell you about the people behind the beverage. "Start with good beans," the primers say. But how do the beans become good? And what does "good" even mean? My micro-prayer of thanks includes those who plant, transplant and tend to coffee shrubs—and the laborers who pick the small red and yellow berries (collectively called "cherry") that surround the coffee seed.

135

But what about the people who transport the harvested cherry to processing stations where the skins of the berries are removed? How about those who ferment, wash, dry, hull, sort, grade and, again, transport the crop? What about the buyers who travel directly to countries of origin to purchase coffee—or the exporters who sell it? Or the people who organize shipments from port to port? Or those who sort, mill, blend and roast the beans once the coffee arrives at its end destination? And how about the clerks who sell coffee in supermarkets—or the baristas who make gorgeous lattes in cafés?

My prayer should be a lot longer.

I pour the ground coffee and water just off the boil into the plastic Aeropress cylinder I've been carrying around the world with me. I stir and—still impatient—wait a minute or two before pressing down the plunger that extracts the coffee directly into my mug. I splash a thimble of milk into the dark brown liquid and finally take a sip. *Now* my day begins. Thoughts come into focus; what was fuzzy suddenly becomes clear. I pull on my clothes and drink coffee, check email and drink coffee, ease—or hurl—into my day, through the grace of coffee.

Coffee used to be my carrier for a cigarette. In the trenches of one of the most stressful love/work periods of my life, coffee and a cigarette were what got me out of bed. They pair beautifully, like wine and cheese, and were the firm beam I steadied myself against as the day unfurled. Coffee and a smoke—one of each, only in the morning—allowed me to simultaneously rev up and calm down, exactly what was required in that bitter Kansas winter when nothing seemed to go as planned. My job was punishing. The relationship I hoped would buoy me up dragged me down; my heart was an anchor. So, each morning, I wrapped myself up in a blanket, stepped outside and drew smoke deep into my lungs as I drank in the sweet warmth of coffee.

Of course it was destructive, but it was also restorative. Those moments were the only ones that felt fully peaceful and solely mine: a time of quiet contemplation when I collected myself before facing the day. The coffee and cigarette were my profane morning meditation: *Slow down, take it all in, breathe.* Those focused five minutes set the tone for my entire day. When I struggled in the afternoon with overwhelm and sadness, I'd start the countdown: *Fifteen hours until the next coffee and cigarette.* Some nights, I went to bed dreaming of the morning.

The longing wasn't just for caffeine. If I wanted that, I could have had tea—or a Red Bull. What I wanted was coffee: the warmth, aroma and taste of the substance I had learned to love.

I drank my first full cup of coffee at the original Krispy Kreme doughnut shop in Winston-Salem, North Carolina. I experienced it in much the same way I received the Bud Light I was discovering around the same time: I hated it. But coffee was far more important to my high school clique than beer. We were intellectuals engaged in deep, post-midnight conversation; we *needed* coffee. I masked the bitter taste with audacious amounts of powdered creamer and sugar, then chased the coffee with a glazed doughnut. The only origin I considered was the one brought to me courtesy of television commercials, ones that told me the best coffee was mountain grown, cultivated in Colombia by a man with a big hat and thick moustache named Juan Valdez.

None of it was particularly interesting to me—neither the coffee nor the farmer. What I was drawn to was what it stood for: a delivery system for caffeine that served as the stimulant for conversation. And I wasn't alone. It's one of the reasons why coffee is now the most widely consumed legal psychotropic drug in the world—a beverage 61 percent of American adults hit on any given day.[1]

Coffee's chemical makeup is similar to adenosine, a molecule our brain produces when fatigued. When we're tired, adenosine attaches to corresponding receptors in our brain that tell the nerve cells in our body it's time to slow down. However, those cells can't tell the difference between adenosine and caffeine (the stimulant in coffee, tea, chocolate and other substances that energizes our central nervous system). So, when we drink coffee, caffeine binds to these same adenosine receptors—but instead of slowing down, our nerve cells perk up. Our brain releases adrenaline into our bloodstream, and alertness ensues.[2] Caffeine stimulates, in much diminished form, the same parts of the brain as cocaine, speed and heroin.[3]

Coffee fuels our day and, as one of the world's most lucrative agricultural crops, fuels economies, accounting for exports worth about $20 billion.[4] The industry "constitutes the livelihood of an estimated 25 million families [and] the worldwide coffee market spans some 71 countries, of which 51 are significant producers and 20 are key consumers."[5] These are the people who help us consume a staggering 1.6 billion cups of coffee—every day.[6] But while almost all coffee is produced in developing and less developed countries, the beverage is consumed mostly in industrialized ones, with the European Union, the United States and Brazil leading the charge.[7]

Coffee isn't just responsible for inciting conversation; it has also played a role in fomenting revolution. "One of the ironies about coffee," says Mark Pendergrast, author of *Uncommon Grounds: The History of Coffee and How It Transformed Our World*, "is it makes people think. It sort of creates egalitarian places—coffeehouses where people can come together—and so the French Revolution and the American Revolution were planned in coffeehouses."[8]

The drink also takes center stage in the cultivation of friendships ("Let's meet for coffee") and discovery of places. For me, that's included having my future read in Turkish coffee grounds in Istanbul and carrying my life in Italy back with me to the United States through espresso brewed in a moka pot. An invitation to coffee is different from an invitation to cocktails; it's comfortable and safe, absent of any pretense of seduction. And it's comforting, which is why the one and only time I got fired, I sought solace not in the bottle, but in traditional Indian cold coffees, one after the other, in Mumbai's Taj Mahal hotel.

I eventually (almost, for the most part) dropped the cigarette, but coffee is still one of the most consistent parts of my life. Relationships change, jobs come and go, but coffee is the constant: the quiet moment, the tone setter, the glorious threshold I cross to enter the day.

Until recently, I didn't know this mainstay was built on the sweat and hopes of people who live continents away, borne of hands that nurture and process fruits the size of cranberries, controlled by a set of decisions that are entirely out of their hands. While I was grateful for coffee in the abstract—for the substance, not the hands that gathered and transformed it—I had no idea how much more there was to learn and appreciate ... or how complicated a simple cup of coffee actually is.

My lesson started on my first morning in Melbourne, Australia. Through a thick fog of jet lag, I raced to Seven Seeds Coffee Roasters, one of the top roasters in one of the undisputed coffee capitals of the world—a logical place to try to understand the extensive work involved in getting coffee from farmers' hands to mine. I got lost and arrived late (because I hadn't had my coffee).

"In the cup, coffee will tell you everything," announced Aaron Wood, former head roaster for Seven Seeds. He was speaking to a small group that had gathered at 9 A.M. on a Wednesday morning for a public coffee tasting, known in the coffee world as a "cupping." He gave a nod and a smile as I stumbled into the small, glassed-in room in the middle of the café. I apologized and explained I'd gone to the wrong coffee shop. "No worries," he replied. His calm demeanor would later carry me through moments when the complexity of coffee seemed overwhelming. "Come in and grab a clipboard." I pulled one from the center of a circular wooden table filled with empty coffee cups, pencils and sheets of paper that the other participants and I would later use to describe and grade the coffees.

Aaron passed around the table grounds we were to sample black—without milk or sugar—and "blind," absent any reference to the types of bean or places where they were grown. As we learned with chocolate, you can't really find the subtleties of flavor when they're hidden behind fat and sugar. The tasting began with smelling: drawing in the scent of the dry grounds. Aaron explained that grinding beans fresh is one of the best and easiest ways to capture the aroma attributes of coffee. He invited us to use our hands to funnel the aromas out of the white cups and into our noses. I cupped my hand and opened my mouth slightly to draw the aroma back into my retronasal passage.

He paused to pour hot water onto the grounds, set a small timer, and picked up where he'd left off: "When you taste a coffee, it reveals its whole story. It will tell you where it's from, who grew it, what variety it is, whether the season was rainy or dry. It will tell you how it was roasted and how it was brewed. It will tell you everything."

I was baffled. How could a simple cup of coffee hold so much?

We leaned in close to smell the wet aroma of the grounds that had risen to the top of the cup, forming a thick crust. Aaron used a spoon to push the grounds to one side ("break the crust") as he, again, inhaled. "The hot water activates all the aromas. We break the crust to release all those volatiles and start to evaluate sweetness." I let my nose hover just above each cup, opened my mouth slightly and inhaled the scent of coffee. Only the scent of coffee. Aaron moved from cup to cup, drawing two metal spoons together to deftly lift out the grounds. We waited for the coffee to cool down (closer to body temperature) to allow for the release of a wider range of the volatile aromas.

Aaron started his journey in coffee at age 26, bagging beans at Atomic Coffee Roasters in New Zealand. "I'd get to work really early," he later told me, "do all my bagging and then clock off and learn from the guy who had roasted in the U.K." Eight years later, Aaron was overseeing the roasting of almost 1,400 kilograms (about 3,000 pounds) of coffee a week. "I love coffee because it's for the people. It's social. You wouldn't go out for a bar of chocolate, would you?" (Thank God, I thought, because I really hate sharing chocolate.) "People drink wine to get out of their day and get into their night. Coffee brings you into your day."

The roasting Aaron oversaw at Seven Seeds is a lot like fermentation—it's alchemic: a process that transforms the subtle aromas and vegetative tastes of green coffee seeds into the shiny, dark beans we crave. Roasting draws out flavors and changes the color of foods through a sequence of chemical reactions between amino acids (the building blocks of protein) and reducing sugars (any carbohydrate that can be oxidized) in what's collectively known as the Maillard reaction. This effect—also referred to as "nonenzymatic browning"—is a major contributor to the mouthwatering appeal of crusty bread, seared steak and golden-brown French fries, as well as the rich flavors and colors we find in coffee, chocolate and beer.[9]

The Maillard reaction was identified just over a century ago by French scientist Louis-Camille Maillard, but was largely ignored because its chemistry was so complicated. The reaction isn't one thing, it's many: thousands of different molecules responsible for hundreds of different flavor compounds and changes in pigmentation.[10] The turning point came in 1953 when African-American chemist John Edward Hodge published a paper that organized Maillard's discovery into three detailed steps.[11]

In coffee, the Maillard reaction is considered "the godfather of all browning reactions."[12] Along with caramelization (another process that occurs during roasting), it reduces bitterness and releases the antioxidant properties of coffee as it changes the aroma and flavor chemistry of the seed into a bean.[13]

"It's elemental in a weird way," Aaron explains. "All we're really doing is making it brown, but it changes everything." A lighter roast can bring out the best qualities of the seed and highlight terroir. Over-roasting, according to Aaron, "can roast all the goodness out but also hide all of the shit in not-nice coffee. The great thing is that it's all part of the most rewarding aspect of the job: transferring great coffee to the customer."

Aaron's modesty aside, a roaster's job shouldn't be underestimated. Kc Reynolds, the woman who trained under Aaron and is now head roaster at Seven Seeds, helped me understand that roasting technique is the equivalent of a signature. Starbucks, for example, roasts all of its coffee really dark (hence, the nickname "Charbucks"), while Seven Seeds strives for a cup that's clean and sweet. "You roast for your location," she says. Meaning, you roast for the people who come to your coffeehouse.

An over-roasted coffee is the equivalent of charring a high-quality steak: You probably don't want to do it because what you then taste is something burnt—the heavy roast, instead of nuances in the bean. In coffee, this can also translate to bitterness, the scale of which, Aaron says, is "exponential."

That exponential bitterness is why some steer clear of the beverage. Evolutionarily speaking, bitter is the taste of poison, something we and other animals understandably shied away from. But now that we're no longer prehistoric beings foraging for food, this dislike exists on a continuum. We have evolved to a point where our genes might hold a dislike of bitter foods—or might not. For one-fourth of the human population who have a taste receptor gene known as TAS2R38, bitter foods are unpalatable. Researcher Valerie Duffy explains, "The idea of how bitter you taste something is [tied to] how strongly the bitter [compounds] in food bind with a receptor. Then the receptor sends a signal to the brain that says, 'Oh, this is bitter.' "[14] And this isn't limited to bitterness: In a comparative study of artificial and natural sugars, researchers at the Monell Chemical Senses Center in Philadelphia found that genetic factors also "account for about 30 percent of the variance in [people's] sweet taste perception."[15]

In other words, we don't all experience the intensity of bitterness—or sweetness—or taste things the same way. For people who don't have bitter taste receptors, those foods and drinks can actually taste *sweet*. But for those who do, an over-roasted coffee is dreadful.[16]

The greatest sensory impact of roasting to the end drinker is in taste, but for the roaster, one of the most influential senses is sound. Aaron, Kc and their roasting colleagues not only monitor the temperature and color of roasting beans, they listen for a series of tiny pops three-quarters of the way through the roast. These pops, which sound a lot like corn popping, are known as "first crack." It's the sound of the release of steam and gasses that build up in the cell walls of beans when they reach a temperature of about 200 degrees Celsius (392 degrees Fahrenheit). The beans start to lose weight and moisture, Kc explains, but puff up as they expand.

After about two minutes, the beans quiet down until the temperature climbs to 446 degrees, setting off another chorus of sounds. The second crack is subtle and sounds more like a bowl of Rice Krispies cereal. After that crack, roasting ceases; the beans are so hot they can potentially catch fire and burn.[17]

As I sat quietly with Kc and listened for crackles and pops, the intimacy of sound became clear. "What we call 'sound' is really an onrushing, cresting, and withdrawing wave of air molecules that begins with the movement of any object, however large or small, and ripples out in all directions," Diane Ackerman explains in A *Natural History of the Senses*. "First something has to move—a tractor, a cricket's wings—that shakes the air molecules all around it, then the molecules next to them begin trembling, too, and so on."[18]

Sound, like touch, is a mechanosensation, meaning it's only activated through a connection to the movement of molecules in the outside world. When it comes to hearing, the ear is mechanosensation's primary mechanism, moving vibrations through three distinct areas to progressively become what we know as sound. Like smell and taste, hearing manifests through nerve impulses that our brain interprets. But the translation is specific to each of us: If a tree falls in the woods and we're not around to hear it, it doesn't make the same sound that we would individually register.

Sound starts with vibrations: mechanical disturbances of air characterized by frequency (cycles of high- or low-pitched sound waves) and intensity (loudness). The waves travel through the air toward the center of our head, starting with the outer ear, called the pinna, and into an air-filled chamber that holds the ossicles, three tiny bones in the middle ear. These bones, the smallest in our body, connect to the eardrum, a thin layer of skin between the middle and inner ear. Sound waves flow down the ear canal, hit the eardrum and cause it to vibrate. Vibrations entering from the ossicles then cause the fluid in the cochlea—the snail-shaped tube that makes up our inner ear—to launch into motion like a wave. Those waves move the roughly 20,000 hair cells lining the cochlea, which turn vibrations into nerve impulses that travel to the brain via the auditory nerve. There—in the brain—they finally become sound.

I didn't know then, as I listened to the *pop! pop! pop!* of roasting coffee beans, that sound affects the way we taste coffee. But it does. Although it's often overlooked in flavor research, sound also informs our sensory experience of foods. What we hear while chewing, slurping, even tearing open a package, shapes our perception of texture and works with other senses to build our expectations and actual experience of flavor. It is so nuanced that "96 percent of people can tell the difference between hot and cold, just by the sound."[19]

Whether we recognize it or not, our experience of flavor is a product of our environment. A 2010 study by researchers Charles Spence and Maya Shankar at the University of Oxford in the United Kingdom asserts, "What we hear influences everything, from what we choose to eat to the total amount and the rate at which we eat it, and to the perceived pleasantness, identity and flavor characteristics of the food in our mouths."[20]

In a follow-up study called "Eating with Our Ears," Spence calls sound the "forgotten flavour sense" and explains, "What we hear can help us to identify the textural properties of what we, or for that matter anyone else, happens to be eating: How crispy, crunchy, or crackly a food is or even how carbonated the cava."[21]

Our sensory experience of food can change up to 60 percent depending on the sounds in the room, such as an ocean soundtrack that makes a fish dinner seem fishier. Researchers at Cornell University found, through their investigation of airline passengers' in-flight beverage preferences, that noisy environments impede our perceptions of sweet tastes but "dramatically increased" our perceptions of umami, or savory, tastes (such as tomato juice).[22]

Increased noise also causes people to rate foods as "less salty, less sweet, yet crunchier."[23] Crunchiness convinces us that some foods are fresher and, possibly, more interesting, leading researchers to suggest that tweaking the sounds of food might be a way to encourage elderly people to eat when their senses of taste and smell diminish.[24]

High-pitched sounds have been shown to make foods taste sweeter, while low-pitched sounds increase bitterness.[25] Sounds of crackling bacon made one bacon-and-egg ice cream whipped up by researchers taste more like bacon, while sounds of farmyard chickens dialed down perceptions of pigginess in that very same dish.[26]

Hearing, of course, isn't the same as listening. In the same way we're always breathing, we're always hearing. But when we pay attention, it's transformed: breathing into inhaling, hearing into listening. Drinking coffee is what many of us do every day, but *tasting* coffee—paying attention to all that is revealed through our senses—is something else, another form of listening entirely.

Aaron dipped a spoon deep into the lukewarm coffee and loudly slurped it into his mouth. I was startled; the sound that emanated from this calm man was piercing, as if he were shrieking. He didn't explain his fierce slurp to the group, so I investigated on my own: "Slurping loudly means you are sucking the coffee in with such velocity that it is aerated and sprayed across your entire palate. This is just like aerating wine—except this time, your mouth is the decanter."[27] That sound isn't necessary, according to Kc, but has nevertheless become an integral part of the ritual of cupping.

Aaron let the liquid rest on his tongue, occasionally drawing it back into his mouth to increase aeration. His blue eyes softened behind his thick-rimmed glasses, and he gazed out over the coffee cups in concentrated meditation. I could almost see his impressions of the coffees percolating up as he discreetly spit the liquid into a paper cup before jotting down his thoughts.

We all followed suit, though my slurp was more of a whisper than a shriek. I sucked in, trying to aerate like Aaron, but I focused less on the sound I was making and more on trying not to inhale coffee into my windpipe. Aaron, meanwhile, moved around the table, sampling coffees and writing down observations: lemon, stone fruit, jasmine.

He then eased off his meditative focus and turned his attention back to us. "What do you reckon?" he asked, sharing a few descriptors while also leaving space for us to discover and share our own. I nodded in agreement to everything he and my fellow cuppers said but added nothing. All I had found in the slurped-up spoon was coffee.

My palate was used to Italian espresso. While I could say with certainty that I preferred a cappuccino to a flat white, I couldn't detect honey, strawberry jam or any of the other flavor notes Aaron rattled off. It was embarrassing. I was writing a book about food; I thought I understood taste. Aaron was sympathetic: "It's okay. It usually takes people a few weeks to really get the hang of this."

The following week I returned, determined to uncover every taste and smell, reminding myself, as I stared at the colorful bits of tattoos peeking out from under the long-sleeved shirt he'd buttoned right up to the top, that Aaron had started out in the same place I did—in the same place we all do: finding one flavor (which, in this case, was coffee). But the intention—the goal—was to cultivate the patience to allow for the possibility of a broader experience.

I came back to Seven Seeds, again and again, eventually memorizing the contours of Aaron's In Love/In Death tattoos. I sniffed, swirled, slurped, spit, searched—and occasionally found. The recognition came about slowly, after nearly a month, testing my patience. When I stopped focusing so intently on one flavor, more showed up. The coffee revealed more of itself to me, just like Aaron said it would.

At the end of one of my final cuppings, Aaron said, "We have a right to know what's in our cup and on our plate, don't you think?" I did. And I wanted to know more.

Specialty coffee, like craft chocolate, is a category that continues to evolve. The term "specialty coffee" was coined by Erna Knutsen during a 1974 interview in the *Tea & Coffee Trade Journal*. Erna, considered one of coffee's living legends, used it to describe unique coffees from Ethiopia, Yemen and Indonesia that had been produced in special microclimates.[28] While in her early 40s, Erna worked as an executive secretary for a San Francisco coffee and spice importer and decided she wanted to cup coffee so she could better describe the flavors to potential buyers. Although she had the blessing of her boss, her male colleagues were less than supportive. "If that cunt comes in here, we're quitting," she described them as saying.[29] But she didn't quit—and they didn't, either.

In 1985, Erna founded Knutsen Coffees and became the first woman to go from secretary to global coffee seller.[30] She later helped launch the Specialty Coffee Association of America (SCAA) and a burgeoning movement to recognize and reward the care that select farmers, roasters and brewers put into their coffee.

Although the specialty coffee industry is small, it's become mighty. Characterized by meticulous care from cherry to cup, it is distinct from coffee sold on the commodities market and, in the United States, has grown from 9 percent of the total coffee market in 1999 to 31 percent in 2015. Commodity coffee sales, meanwhile, have remained flat.

Specialty coffee is, in a nutshell, "coffee that's special," explains Peter Giuliano, one of my coffee mentors and senior director of the SCAA. The sector is vital to the sustenance of the entire coffee industry because it holds not only the hope of increased deliciousness, but also access to the more diverse varieties needed to sustain the crop. But deciding where and how a coffee becomes a *specialty* coffee isn't an exact science. "In green coffee?" Peter asks. "In roasted coffee? In the cup of coffee? It's a multifaceted question. At the very least, it has to mean an absence of bad things and presence of good things."

Historically, the starting point for the "absence of bad things" was in physical defects in unroasted, green coffee. As Peter explains, "It's the ways the coffee seed has been physically compromised by insects, bacteria, chemicals and the like." Or, as Aaron said during cuppings, "no mold, no earthy notes, no astringency ... coffee that is picked when ripe and has all the rubbish sorted out."

Through the 1980s and 1990s, the SCAA developed a cupping protocol that didn't just explore defects but also evaluated the presence of positive flavor attributes. Peter explains: "Specialty coffee has expanded from the original reference to one that highlights specialness, from the region to the farmer, and from the roaster to the barista."

At the end of each blind cupping, coffees are scored based on defects (which subtract points) and attributes (which add them). "If the coffee scores less than 80 points," Peter says, "it's not specialty; if it's above 80, it is. And above 90, things get really great."

Seven Seeds uses 85 as its magic number. It only purchases coffees that cup above that score, ones that allow for a very small number of visual defects, "and definitely no defects you can taste," Aaron adds. "No processing defects like mold or ferment, no overripes or underripes, no insect damage or disease defects. And no baggy coffee. Ever."

Although it's quantified, the decision to purchase—or not—isn't easy. Aaron described for me relationships he and his colleagues had built in Honduras, working with farmers for over a year to improve their growing and post-harvest processes. "But then the following year, it cupped at 82 and we had to tell them we couldn't take it. It was really, really tough. Even at a company that seemed to care, I was told the price had to be right and the 'cup score' had to be right. I just wanted the coffee to be right. It's people, it's children and it's environment. I couldn't do enough."

Rejected shipments are eventually sold because, like cacao, there's always a market for coffee—though not likely at the price the farmers were counting on, and not likely in the form the market intended. Rejected lots, and any special flavors they may have held, are blended into bigger parcels of coffee and roasted until they all taste the same. Coffees with the most defects tend to stay in the country of origin or get shipped to poorer countries.

"What we're ultimately looking for," Aaron explained, "is a coffee that's clean and sweet" (the signature flavor of Seven Seeds). "It's tricky, because every process along the way is going to affect that." By "clean," Aaron means free of defects, and by "sweet," he means just that. "For me, sweetness is a yes or no answer. If the coffee is harvested, processed, stored and shipped well—and free of defects—it will be sweet." But that sweetness has to be balanced: "No one wants to eat fruit or drink wine without acidity; it would be flat and dull—like a flat soda."

Until my forays into caffeine, I assumed acidity was a negative attribute, associated with acid reflux and a reason people gave up coffee. But what coffee professionals mean when they talk about acidity is brightness: a slightly tangy lift we find in fruits, one that makes our mouth water. "You could almost hinge everything off sweetness and acid," Aaron says. "A citric acidity will remind you of mandarin oranges and tangerines. Malic acidity will remind you of red and green apples, but acetic acidity—you don't want that. That's vinegar."

This acidity is also found in cacao (and is the quality TCHO highlights in their "bright" bars), as well as in other foods and drinks. But for those accustomed to instant coffee or espresso-based drinks, it can come as a surprise. I loved the glass of Trousseau Gris Ali brought me from the first moment I tasted it because it was unexpected but still reminiscent of what I knew. That was *not* the case with my first foray into Ethiopian coffees. They had notes of lemon I had never before experienced in coffee and were so far from what I knew, I thought the milk I had added to my drink was sour.

When I admitted this to Aaron, his eyes practically danced: "We think it doesn't taste like coffee because 95 percent of what we drink comes from just a handful of coffee varieties. Not diverse, heritage plants like Geisha or some of these other heirlooms but something typical and one-dimensional. I mean, when they're grown or processed well, those varieties can also be great, but coffee—it's not one thing." Differences in variety aren't the only factors that influence flavor (environmental and management practices are also contributors), but they create a range I had not experienced until that moment.

Coffee is from the Rubiaceae family of flowering plants, which includes ornamental gardenias, *Rubia* plants used for dyes, and *Cinchona*, the bark of which contains quinine and is used to treat fevers and malaria. Although there are about 125 species in the *Coffea* genus,[31] we basically drink only two: *Coffea canephora* and *Coffea arabica*.

Instant churn-and-burn is made mostly from *Coffea canephora*, more commonly known as robusta. Robusta lives up to its name: It has greater resistance to pests and diseases and is less vulnerable to climate change. Its toughness is also reflected in the coffee's strong, woody taste, increased bitterness and higher caffeine content, which is why it's sometimes mixed with arabica to turbocharge espresso blends.

Along with a species known as *Coffea eugenioides*, robusta is the parent of *Coffea arabica*, the hybrid that makes up the majority of coffee we drink. "'It's a love story actually,'" explains Timothy Schilling, executive director of the industry-funded, nonprofit institute World Coffee Research. "'Arabica has two parents that met some 10–15,000 years ago and combined to create Arabica [sic]. It was a one-time-only event, a one-night stand, if you will.'"[32]

Most of the hybrid matches we see in modern-day agriculture are engineered by strategic crossbreeding and designed to manifest specific traits (such as the disease resistance and increased productivity bred into the cacao hybrid CCN-51). But the rendezvous that birthed arabica occurred naturally, over time.

Arabica is a bit precious: It has a more refined flavor and greater susceptibility to temperature fluctuations and diseases. It makes up the tins of French Roast lining shelves at Trader Joe's, the espresso in a Starbucks latte and the more exotic types of coffee we find in cafés, including Caturra from Brazil, Sumatra from Indonesia and Blue Mountain from Jamaica.

Yet, despite the vast geographical range of cultivation, the number of cultivated varieties (known as cultivars) of arabica that are grown for commercial production are limited. There are hundreds of varieties in the arabica species,[33] but we mainly consume Typica (the oldest variety), Bourbon (a natural mutation of Typica that occurred on the island of Bourbon), and hybrids of the two.[34]

Every time I cupped at Seven Seeds, Aaron, Kc and their colleague Matt Ledingham would try to share something new, something that would help me better understand that coffee wasn't one thing, but many. They had each experienced a moment that transformed them—and they wanted the same for me. Matt told me about the first time he tasted Geisha from Hacienda La Esmeralda, an heirloom coffee from Ethiopia now grown in Panama. "It was phenomenal," he said. "I don't want to sound like a wanker, but it was. It was about becoming conscious of something—that coffee as I knew it or understood it could hold those kinds of flavors. And saying, 'Yes, I love this in a new way.' It was an illuminating moment."

For me, that moment was quiet. It happened in my Melbourne apartment with a coffee from Yirgacheffe, Ethiopia, that my girlfriend Jeni had included as part of a care package she'd sent me from California. (The coffee had been roasted at Sightglass, an independent roaster in San Francisco.) At that point, I had grown used to coffee tasting like more than coffee, but what I didn't expect were flowers. That coffee did what the wine did: When I least expected it, it reached for me. Jasmine and honeysuckle bloomed in my cup. I wanted to start every single morning with that caffeinated bouquet.

The story in that—and every—cup of coffee starts in Africa, the cradle of civilization, the region where we learned to forge tools, tame fire and develop language. Every time we drink coffee, we're consuming an African crop. In my early conversations with Peter, he explained, "Coffee is known for tasting different from place to place, but there is more flavor diversity within Ethiopia than in all other countries of the world *combined*. Every flavor coffee is capable of producing can be found in its birthplace." That's because all of the diversity of arabica coffee is found there: It's the origin of all coffee origins. Along with the forests of South Sudan, Ethiopia is the center of origin and diversity for the majority of coffee we drink.

Ethiopia is also one of the world's 34 biodiversity hotspots: the center of origin and diversity for not only coffee, but nearly 40 crops, including barley, millet, durum wheat, black-eyed peas (cowpeas) and sesame.[35] Every sip of coffee connects us to the cradle of civilization, to the place where we evolved into early humans and became who we are.

Legend has it that, in Ethiopia's southwestern highlands, a goatherd named Kaldi saw his flock perk up after eating bright red berries off a leafy green shrub. Kaldi partook—and the story of coffee began.

Around the year 525, Ethiopians invaded southern Arabia. They ruled Yemen for about 50 years, during which time they cultivated coffee.[36] In 1536, the Turks conquered Yemen and folded coffee culture into their own. Issues that couldn't be addressed in mosques were discussed in public spaces—over coffee. While Europeans were starting their days with beer, Ethiopians and citizens of the Arab world were fueling their lives with caffeine.[37]

Coffee was a major source of income for the Ottoman Empire. To protect its monopoly on cultivation, the seeds were parboiled or partially roasted before being transported. But this changed when a Muslim cleric named Baba Budan—amid his pilgrimage to Mecca, in a move that was decidedly *unholy*—smuggled seven viable seeds out of the country to India (the story behind the name Seven Seeds). Indian coffee trees were the first ones cultivated outside of Africa, effectively breaking the economic stranglehold on the crop.

Just like the Kerman nut that established the U.S. pistachio industry, most of the coffee cultivated around the world was established with a single shrub. Coffee moved from Ethiopia to Yemen and then on to India and Indonesia. "And it went from there to a greenhouse in

Europe," Peter explains. "And one plant [from that greenhouse] was taken ... to Martinique. That one plant became the forefather of all the coffee that was grown in Latin America."[38]

Coffee became a leading crop in Ceylon (what's now Sri Lanka) and in Brazil, where it was introduced in 1727, seeding what has become the biggest coffee-producing country in the world.[39]

Those first plants not only transformed what was grown in producing countries but also reshaped what people drank throughout Europe, leading Pope Clement VIII to exclaim in 1600, "Why, this Satan's drink is so delicious that it would be a pity to let the infidels have exclusive use of it. We shall fool Satan by baptizing it and making it a truly Christian beverage."[40]

Matthew Green, author of *The Lost World of the London Coffeehouse*, explained that, until the mid-17th century, "most people in England were either slightly—or very—drunk all of the time. ... The arrival of coffee, then, triggered a dawn of sobriety that laid the foundations for truly spectacular economic growth in the decades that followed as people thought clearly for the first time. The stock exchange, insurance industry, and auctioneering: all burst into life in 17th-century coffeehouses ... spawning the credit, security, and markets that facilitated the dramatic expansion of Britain's network of global trade in Asia, Africa, and America."[41]

Coffee didn't just launch markets, it sparked political discourse and activism: "Traditionally, informed political debate had been the preserve of the social elite. But in the coffeehouse it was anyone's business—that is, anyone who could afford the measly one-penny entrance fee."[42]

Coffee consumption became a method of protest against British manipulation of tea prices in colonial America in defiance of the Tea Act, which the British Parliament passed in 1773 to ship cheap tea directly to the colonies, undercutting prices offered by local businesses.[43] Following the Boston Tea Party, United States founding father John Adams wrote a letter to his wife, Abigail: "Tea must be universally renounced and I must be weaned, and the sooner the better."[44] Drinking coffee had become an act of patriotism.

During the 19th century, most coffee-drinking Americans roasted their own beans, brewed strong cups of coffee and clarified the grounds by adding dried fish skin or eggshells.[45] That intense brew powered factory workers during the Industrial Revolution, empowered soldiers on the battlefield—some of whom had coffee mills built right into the buttstocks of their rifles[46]—and, as coffee moved from the trenches of war to the kitchens of suburban housewives,[47] became entrenched as a quintessentially American way to start the day.

The most popular birth story of coffee culture (beyond the crop) starts with the rise of the drink in European cafés: milky Viennese coffees sprinkled with chocolate flakes and Italian cappuccini topped with creamy tan peaks that approximated Capuchin monks' brown, hooded robes. But the roots of coffee consumption also belong to its birthplace: Ethiopia is the center of origin for the cultivation and consumption practices that gave rise to our contemporary coffee culture. Despite the worldwide popularity of coffee, Ethiopia remains its past, present—and future.

When I landed in Addis Ababa, I already knew it was like no other place I had visited. The country has a calendar with 13 months and marks the start of a new day at sunrise, not midnight. (This resulted in a moment of extreme panic when I realized I hadn't clarified if my meetings with farmers and scientists were in East Africa Time or local time.) Ethiopia has never been colonized and twice drove out Italy, the only country that ever tried to occupy it. The only remnants of the failed occupations seem to be terrific Italian food and the prolific use of the word "ciao." Ethiopia belongs to no one but itself.

And, like coffee—like *everything*—the country isn't one thing, but many. It's lush and arid; the sun is bright and shadows are long; people are thin and wide—and generous beyond measure, in ways they don't have to be. The opening question everywhere, with everyone, is one of interconnection: not just "How are you?" but "How is your family?" Family is the thread that ties everyone together. This constant query, from every person I met, reminded me that I was connected to others. But what charmed me most was the way people register agreement. In

Ethiopia, the equivalent of a quick head nod or a verbal "uh-huh" is a sharp but soft intake of breath—a light but audible gasp.

The country took my breath away.

Ethiopia is one of the world's largest coffee producers, with sales of *bunna* (coffee) accounting for more than 10 percent of government revenue and 25 percent of the country's earnings from exports.[48] It is the only African country that prefers coffee to tea. So, as delicious as a brew from Kenya or Rwanda might be, for those nations, coffee is just a cash crop. For Ethiopia, it's lifeblood: fuel of the culture and economy.

Over half of the coffee grown in the country stays there, and because everyone knows someone involved in coffee, expertise and exposure are democratized. This is in stark contrast to the West, where coffee is consumed most widely, yet knowledge about coffee stays in the hands of an expert few. No matter where you go in Ethiopia—no matter whom you meet—everyone has something to say about coffee: the driver, the waitress, the man who issues your visa at the airport. One in four Ethiopians is directly or indirectly connected to coffee as a source of income.

"Coffee is everything," explains Frehiwot Getahun, manager of the Kafa Forest Coffee Farmers Cooperative Union. "It is identity. It is employment—not just for the farmers but for the managers, laborers, administrators, truck drivers, cuppers, traders. And it is like food. You have it in the morning, the afternoon, the evening. People live on, and through, coffee."

Ethiopia taught me that the greatest coffee experts don't only hail from Seven Seeds or live in Portland or Oslo. The most experienced people in coffee—ones who have grown up with an understanding of our favorite brew from soil to cup—are the girls and women I watched repeatedly pound pestle into mortar.

"A daily dose of *bunna*," ethnographer Metasebia Yoseph explained in her book A *Culture of Coffee*, "is more than just a caffeine fix; it is a cultural fixture."[49] Every household is a coffeehouse, and the head roaster and master barista is a woman who starts her training at age 10.

Hana, a 13-year-old girl from Addis Ababa, worked in the home of the family with whom I stayed when I first arrived. She was slightly less knowledgeable than her compatriots in rural parts of the country where coffee is grown, but far more expert than I would ever be with my hand mill and Aeropress. Under the watchful eye of Alemitu Tilahun, the matriarch of the family, Hana performed the traditional coffee ceremony, a ritual that has been passed down for generations.

The coffee ceremony is the opposite of my solitary morning ritual of coffee and a cigarette; it's an opportunity for connection. Neighbors are called, frankincense is burned and green beans the color of sea foam are passed around for everyone to inspect before they are placed on a cast-iron griddle. Together, we sit, we wait—and we engage.

Hana's setup was a flat pan resting on coals, tended carefully as her single-origin coffee made its transition from living seed to roasted bean. The air was fragrant with incense, the conversation punctuated by the scraping of a metal hook that moved the beans across the skillet. The beans crackled as they toasted to a deep, shiny brown, taking me back to my days at Seven Seeds. *Pop, pop, pop:* The smell of coffee filled the room.

After the beans cooled slightly, Hana pulverized them with her *mukecha* and *zenezena* (mortar and pestle). I crouched down beside her and watched her pudgy hands lift the pestle chest high and *thump, thump, thump* up and down in rhythmic repetition. Although I tried repeatedly to catch her eye, she was absorbed in her task. One hand grasped the metal vessel, the other pounded the roasted beans into a coarse brown powder akin to the soil in which they were grown.

Mama Tilahun then took the grounds from Hana and blended the coffee powder with boiling water in her *jebena,* a clay coffee pot that rested on burning coals. The pot was worn and the spout was broken; it had been used well and often.

From that pot, we were served three times. The first pour, known as *abole,* was dark and strong, a jolt of caffeine that animated our conversation. Hana refilled the *jebena* with water and set it back on the coals to boil as we waited for the slightly weaker second pour, called *tona.* The third and final pour—*bereka*—translates as "blessing."

At home, I had pondered the merits of individual pour-overs—and the ecological demerits of individual K-cups—but that concept never arose in Ethiopia because the coffee break was dedicated to communion. Every time I tipped the cup to my mouth I thought, "Of course coffee evolved in the same place as the human species. *Of course.*" We were sent into our day with a blessing.

The beans Hana pounded into powder came from a roadside stand where local farmers sold coffee that had been grown on their small garden plots, as most coffee in Ethiopia is grown. The designation was simply *bunna:* coffee. The seeds may have been passed on from the local agricultural extension or another farmer, or saved from another harvest. Regardless of the source, the diversity within those seeds has progressively diminished as researchers and farmers have selected crops exhibiting the small number of traits needed to sustain production.

As Aaron and Peter explained, global production of arabica (and robusta) depends on just a handful of cultivated varieties, with little difference at the genetic level or in their physical (phenotypic) characteristics. This is a problem for all the reasons we now know: Reduced diversity equals increased risk.

Coffee trees grow best between 19 and 25 degrees Celsius (66 and 77 degrees Fahrenheit). The productivity of arabica, as Tadesse Woldemariam Gole from the Ethiopian Coffee Forest Forum (ECFF) and Aaron Davis of Kew Royal Botanic Gardens explain in their study on coffee and climate change, is "tightly linked to climatic variability, and is thus strongly influenced by natural climatic oscillations."[50]

Since 1960, the average temperature in Ethiopia has increased by 1.3 degrees Celsius (2.3 degrees Fahrenheit).[51] Drought and erratic rainfall have severely compromised coffee production in the southern part of the country.[52] The climate modeling done by Kew Gardens and ECFF estimates that, as a result of a warming planet, the areas that contain the highest concentration of coffee diversity could be reduced by 65 to nearly 100 percent by 2080.[53] Not only would the country that gave the world coffee no longer be able to produce it, but the diversity we need to be able to access it would also be lost.

Emerging data from World Coffee Research on the diversity of one of the most important *ex situ* collections of arabica coffee, housed at CATIE (Centro Agronómico Tropical de Investigación y Enseñanza) in Costa Rica, shows that Ethiopian coffee accessions collected in the 1950s and 1960s were shockingly similar.[54] Scientist Tim Schilling explained, "We were extremely surprised that there was 98.8 percent similarity. We knew diversity would be low, but we didn't expect it to be nearly absent." The research teams that collected the samples were looking for traditional varieties, not necessarily all the diversity found in the forest—so this assessment might be a reflection of limited sampling or might indicate there's a lot less diversity in the forest than expected.

"As scientists, we can't say the forests of Ethiopia are equally or even close to equally constrained," Tim said. "It's possible that there is more diversity than was revealed. This study is not comprehensive enough to make a judgment." While this hangs in the balance, what we do know is that the world is getting hotter and, if we are only to rely on backup collections of arabica coffee, we're in trouble. Especially when we consider the fact that, globally, coffee plantations contain less than 1 percent of the genetic diversity found in Ethiopia's coffee forests.[55]

This is why the diversity found in the wild is so valuable. Without it, the entire crop could disappear.

II. THE WILD

The forest is where the wild variants of arabica thrive, a place where every plant is fighting for its survival: reaching for the sun through the shade, battling competing vegetation for nourishment and space. These wild plants don't have water and nutrients delivered to them. They don't get doused with chemicals to ward off insects. They're tough because they have to be. The coffee most resistant to diseases and pests, and most likely to tolerate any other changes we'll experience, grows here, in the forest.

The best way to save coffee is to save what's wild.

Arabica coffee is a naturally occurring shrub of the Afromontane rainforest, an evergreen mountain forest in the southwestern highlands of Kafa, where the adventurous goatherd Kaldi allegedly launched our love of coffee.[56] Officially known as the Kafa Biosphere Reserve, it is one of three UNESCO biosphere reserves in the country and "the first coffee biosphere reserve in the world."[57]

My guide was Mesfin Tekle, project coordinator for the Nature and Biodiversity Conservation Union (NABU) Climate and Forest Project. With degrees in both agriculture and natural resource management, Mesfin works to increase the visibility of the Kafa forest and the wild coffee it contains. Coffee is considered wild, he explained, as long as it grows in its natural habitat, without human intervention, and is genetically different from known cultivars. In order to keep the original genetic makeup of wild coffee pure—and to prevent it from interbreeding with modern varieties (which would potentially reduce its genetic diversity)—we have to keep its wild habitat intact.

A no-nonsense man, Mesfin walks fast and talks fast. His brow maintains a constant furrow, but when he gets excited, his eyes gleam and a fleeting, joyous smile crosses his face. At 5 feet tall, it's unusual for me to be able to gaze into any adult's eyes while standing, but Mesfin is 4 feet 11 inches. His eyelashes are the longest I have ever seen.

As we wandered through the wildness, Mesfin explained there are 5,000 variants of coffee within this ecosystem. This includes recognizable coffee plants, plus their wild relatives. "In Ethiopia," he said, "where there is a forest, there is coffee." In place (*in situ*) conservation in the wild, he added, is the most resilient and responsive backup to what's being lost.

Although it was warm and humid in the surrounding area, the minute we stepped under the thick shade canopy, it felt cool. Every surface—the plants, the trees, the flowers, the ground—seemed to pulse with life. Even the air felt abundant: rich with the scent of green moss and damp earth, thick with the drone of honeybees. (Ethiopia contains the largest population of these pollinators on the continent.)

In this wild, verdant place, it was easy to recognize the forest as our deepest reservoir of diversity. What was hard for me to get my head around was why we seem determined to destroy it.

Forests are home to at least 80 percent of the world's remaining biodiversity.[58] While the rainforest once covered 40 percent of the surface area of Ethiopia, by the early 1990s, this area had dwindled to less than 3 percent. According to the United Nations, Ethiopia could be "completely deforested by 2020."[59]

The causes of deforestation are multiple but are largely the result of agriculture, timber extraction, overgrazing and the expansion of communities. These losses aren't just about coffee, of course, but all the organisms that are part of the forest, many of which are disappearing before they've been fully documented or understood.

Deforestation is why we are losing one of the most vital strands of our food web: pollinators—the insects or animals that transfer pollen from one plant into the female flower of another. Author Dave Goulson explains in his book *A Sting in the Tale: My Adventures with Bumblebees* that, without the help of these "sex facilitators," plants are quite inefficient sexual partners, squandering up to 99.99 percent of pollen.[60]

In the Kafa reserve, bees aren't just sex facilitators, but laborers of love. Coffee plants are self-pollinating—meaning they don't need a carrier, such as the wind, a bird or a bee, in order to reproduce—but bees still sip coffee blossom nectar and have a profound impact on the crop. David Roubik of the Smithsonian Tropical Research Institute in Balboa, Panama, found coffee cherry yields skyrocketed nearly 60 percent when the shrubs' flowers were visited by African honeybees.[61] And an additional study in Tanzania found that, when birds and bees spent quality time with coffee blossoms, the cherries were approximately 7 percent heavier.[62]

Yet these small miracles are dying off for reasons we can't fully explain. There are more than 25,000 different bee species in the world, but many are threatened.[63] How many? We don't exactly know, because it's tough to study bees on a global scale. A 2015 study from the International Union for Conservation of Nature offers a sobering snapshot of this biodiversity loss: Nearly 10 percent of Europe's almost 2,000 wild bee species are at risk of extinction, while an additional 5 percent will likely be threatened in the near future.

Even more troubling was the study's assertion that it is impossible to determine whether the majority of European bee species are at risk because of "an alarming lack of expertise and resources."[64] We don't know what we don't know, so it's difficult to fully quantify what we're actually losing. What we *do* know is that the United States is experiencing the second largest die-off since 2010, and beekeepers lost more than 40 percent of their honeybee colonies from April 2014 to April 2015.[65]

Scientists attribute the global decline of pollinators to stress and increasingly inhospitable environments due to industrialized agriculture's use of products like neonicotinoids, a class of synthetic insecticides that have a similar composition to nicotine, the chemical found in cigarettes.

"Like nicotine," explains plant biologist Ken Thompson, "neonicotinoids are extremely effective nerve poisons, but unlike nicotine they are really only toxic to insects and are very safe to use."[66]

Or so we thought.

Swedish scientists studied the impacts of neonicotinoids on wild bee populations and found that "such insecticidal use can pose a substantial risk to wild bees in agricultural landscapes, and the contribution of pesticides to the global decline of wild bees may have been underestimated."[67] Although the study focuses on wild bees, bees all around the world are experiencing adverse effects due to the increased use of these insecticides.

Thompson goes on to explain that the real problem is repeated exposure: A single bee might repeatedly visit a field treated with neonicotinoids. "The sublethal effects are more interesting," he writes. "We're talking here about a bee's ability to perform a variety of tasks, including learning and remembering the location of good nectar sources, and its ability to return successfully from a remote nectar source. ... A bee that loses its way is a dead bee."[68]

And a field without bees is a dead field. That's the thought that came to me at the only coffee plantation I visited in Ethiopia. I could hear people, the occasional chirping of birds and the faint sound of plants being dug out of the earth and transplanted to another site, but little else. The absence of buzzing was deafening.

Wherever Mesfin and I went in the forest, we heard bees. That drone is a soundtrack I can conjure over two years after my visit. I close my eyes and see myself kneeling before plant after plant, surrounded by humming pollinators, trying to capture on my cell phone the perfect image of coffee seedlings. Unlike cacao pods, baby coffee was exactly what I had envisioned: a small, light green shoot erupting from a cracked seed. Alongside the seedlings were coffee blossoms that seemed to be scattered everywhere, like confetti after a party. The tiny white flowers nestled against dark green leaves in an area shaded by old growth trees. The majestic trees that towered high above only exist here in the wild. The living archive held so much serendipity and diversity, it felt holy.

Environmental economists have tried to slow the pace of deforestation by putting a price on "ecosystem services"—a valuation of the wealth of resources various components of our ecosystem provide. But, as Peter Raven, one of the key people responsible for putting biodiversity concerns into public consciousness, explains, "the world total of ecosystem services—the things that are provided by Nature—is *everything*. It has infinite value. If there were no services being provided by Nature, we'd all be dead."[69]

When it comes to coffee, a 2006 estimate of Ethiopian coffee plants by economists Lars Hein and Franz Gatzweiler placed coffee's value somewhere between $420 million and $1.4 billion.[70] This wide range is due to the fact that we don't know exactly what will happen—or what wild and diverse resources we'll need in the future. But while these figures exist on paper, the value of the biodiversity of the coffee forest isn't being embraced on the ground (in the actual forest).

Vegetation ecologist Feyera Senbeta from the Environment and Coffee Forest Forum offers a partial explanation, "When people are cultivating a lot of coffee, they degrade the forest. This is because when the forest is dense, the competition for resources is very intense. Farmers try to reduce competition by removing all small trees and some stumps. But then big trees and small coffee are all that remains."

Put more broadly, industrialized agriculture—the practice of growing food and the means by which we grow the majority of what we eat and drink—threatens the very existence of those foods and drinks. The way we grow food compromises our ability to eat.

Another bitter irony is that one of the main causes of genetic erosion is the steady replacement of wild coffee and more diverse cultivars with a smaller grouping, generally known as "improved varieties"— coffee primarily bred for increased yields, climate adaptation and resistance to diseases like coffee berry disease and coffee leaf rust. It's the same genetic erosion we learned about in Ecuador with CCN-51 cacao. Diversity disappears, bean by bean.

Coffee, like cacao, is an orphan crop, meaning it's under-researched and under-funded. " 'Unlike many other crop species,' " coffee researcher Timothy Schilling explained to BBC News, " 'coffee has had very little research behind it.' Coffee only has about 40 plant breeders, compared to thousands in crops like corn, rice or wheat."[71]

Mesfin helped put this in context: "We have all these different types of coffee, but the world only cultivates about fifty." I gasped. This was a moment when I longed for less complexity, for one solution that would make these problems better and improve the lives of all the people fighting to save what I loved—one that would honor the lineage of all who had cultivated coffee, one that would preserve and sustain its stories. And then I recalled author Bill McKibben's words: "There are no silver bullets, only silver buckshot."[72] The problems—and solutions—are multiple. Every decision comes with costs and benefits.

Coffee leaf rust is the fungus responsible for wiping out huge swaths of Central American coffee in one of the worst outbreaks we've ever seen: "Spurred by unusually high rainfall over the last few years, [the fungus threatened] to ruin as much as 40 percent of the 2013-14 Central American harvest. To appreciate the potential outcome of this threat, consider that the only reason Ceylon *tea* exists is because Coffee Leaf Rust comprehensively destroyed [the country's] once lush coffee plantations in the 1860s."[73] Known in Spanish as *la roya* ("the rust"), it is why Costa Rica, Guatemala and Honduras have declared national states of emergency.

One of the reasons the fungus has taken such hold in these countries is because the coffee plants are now grown in monoculture plantations under full sun. "'Exposing coffee to sun is like giving the plants a shot of steroids. The sun speeds up photosynthesis, accelerating the plant's growth,' says Paul Katzeff, whose company, Thanksgiving Coffee, sells shade-grown Song Bird Coffee. Juiced up, the hungry plant quickly depletes the soil, so that farmers have to apply chemical fertilizers to support the plants. All this activity exhausts the coffee plants. 'It's living in the fast lane, so the plant lives only fifteen years as opposed to eighty or a hundred.'"[74]

To add to the confusion, this change in growing methods and coffee cultivars is an attempt to *solve* the problem. "To prevent the spread of the disease, coffee farmers 'technified,' replacing older, shade-loving coffee varieties, such as Typica and Bourbon, with new varieties, packed in tight hedgerows, that can survive open sun—but only with chemical inputs."[75] They are now so dependent on these chemicals, they—the farmers, the plants, the entire industry—can't survive without them.

Ric Rhinehart, the executive director of the Specialty Coffee Association of America (SCAA), explained to *The Atlantic* that small-scale farmers who grow coffee crops organically (without synthetic chemicals) have suffered the most: "'The best possible solution is an application of [synthetic] fungicides.' Such fungicides are banned by organic standards. Only 3 percent of the crops in Guatemala are rust-resistant varieties. The rainy season is fast approaching. And international coffee prices are at historic lows. 'To put a colorful spin on it,' Rhinehart says, 'these guys are just fucked.'"[76]

This awareness humbles me. In my demand for organic coffee—one that I considered virtuous because it meant less synthetic chemical exposure to the farmers, the ecosystem and me—I hadn't considered what such an absolute position might mean for the farmers who suffer the greatest risk in the supply chain of coffee.

I still want coffee grown with as few synthetic chemicals as possible—for all the reasons I just mentioned—but coffee taught me to soften my stance and allow for some flexibility. "There's a romance that comes along with that, especially in the globalized north—we yearn for virtuosity," SCAA director Peter Giuliano says. "But if you spend enough time with farmers, they want some of those things: A mechanical picker would be great; fungicide application might mean they can save their crop and feed their kids."

Everyone wants a good life. In order to achieve it, we all have to make trade-offs.

In the past 30 years, the southwestern highlands have lost 60 percent of their forest cover. Farmers have cleared densely populated forest to switch to other cash crops. But, as Mesfin explained, "these farmers know nothing but coffee. They have grown little else for generations. Now the returns are so low they can't survive."

The only form of irrigation in Ethiopia is rain. A three-month dry season is all the thirsty coffee plants can bear. With climate change, this dry season grows longer and hotter, putting at risk not only plants but the entire coffee industry: farmers, managers, laborers, administrators, truck drivers, cuppers, traders—the one in four Ethiopians connected to coffee.

In Ethiopia, coffee is grown in four ways. Plantation coffee is grown on larger estates as a single monocrop in full sun and is not a major form of cultivation (as it is in Brazil or Vietnam).[77] Most coffee is garden coffee, grown on plots of fewer than 4 acres, interspersed with other crops.

After garden coffee, the most popular coffee is semi-forest coffee, which can and has been sold as wild (or forest) coffee, but is actually grown on land filled in with coffee plants from other areas. In this method of cultivation, farmers selectively prune coffee shrubs and thin other trees to let in sunlight, eliminate competing plants and create room for more coffee.[78] But it can easily become a slippery slope where lots of trees are cut down—in one forest 30 percent of vegetation was removed[79]—or too much coffee is harvested. "The practice of swapping and transplanting coffee, and any interbreeding, erodes the integrity of wild genotypes"[80]— compromising the purity of what's wild and eroding biodiversity.

This is why the fraction of coffee (5 percent) that is *truly* wild is so important.[81] Protected areas, like the Kafa Biosphere Reserve, operate in concentric circles: The bull's-eye area remains untouched, with no human intervention, while the buffer zone allows for harvesting and minimal management.[82]

The farmers who collect wild forest coffee deserve a lot more attention. If they stop picking coffee from these forests, those areas lose their greatest economic value, which puts them at increased risk of being cut down and replaced with other crops, or developed in other ways. The place where coffee was born—the area with the greatest biodiversity of coffee anywhere in the world—could disappear.

No forest, no coffee. No coffee, no forest. What we lose isn't specific to Ethiopia; it impacts us all.

So how do we save the forest and the coffee, and support the farmers who are growing it? Mesfin believes by drinking it. "We have to put more value in these reserves so we can protect them," he told me as we took our second walk through the forest. "Wild coffee is a gift from nature. The cherries ripen more slowly because of the shade cover, so the delicious flavors that are inherent in the coffee are allowed to develop. The taste of forest coffee is strong, smooth and round. It lasts a long time. And it tastes like the place from which it comes—the earth."

Farmer advocate and author of the CRS Coffeelands website Michael Sheridan explains that these tastes aren't the result of the forest but are enhanced by it: "Those thousands of wild varieties in Ethiopia are good because the extraordinary flavor profiles few have ever tasted are baked into their DNA."

But not everyone agrees that inherent earthiness is a good attribute. The mark of specialty coffee is a cup that's *absent* of defects. Wild coffee tastes exactly how it sounds: unpredictable, muddy in the cup and usually sun-dried in its skin (known as the "natural" process), handled without extensive effort because the market is small and not really profitable.

In the chocolate chapter, we learned that not all is as it seems. Because the market is young and people are still defining what "good" means, some cocoa beans that are defective are now being sold as chocolate of high quality. In coffee, we run into a different but related problem: Good flavor has been clearly defined—and most forest coffee doesn't qualify.

But what if we allowed for a muddy cup and embraced the earthiness of the forest? Would it reward bad processing practices and mediocre tastes? Is forest coffee the equivalent of a smoky chocolate bar? Everyone involved with coffee is wrestling with this tension, wondering how to improve processing on a crop that isn't yet fully valued.

When I discussed this challenge with roaster Aaron Wood, he said, "I have had it in my head that the Earth tastes bad. I mean, 'earthy' is considered a defect in specialty coffee. But why the fuck is the Earth a defect? It's a defect because we've made it one. The Earth should taste *amazing*, the Earth should taste fresh, complex and spark in us a connection to it. But we have to love it and look after it."

Perhaps this is the place to embrace the economic edict of letting the market decide. We are the market. We set the standard. What do *we* want? How does our end decision impact what we drink, what farmers grow and gather, and what choices we'll have in the future? This isn't a single decision; it's an evolving step on the journey, as dynamic as the seed from which it came.

The next morning, upon my verging-on-belligerent request, Mesfin agreed to help me get a taste of forest coffee. I had been staying in Bonga, the one-hotel town (if you could call Coffeeland a hotel) situated closest to the Kafa reserve. Although the region is considered the birthplace of coffee, Bonga doesn't seem to recognize its global significance or understand how precious it is. Like the hotel, it feels run-down and tired.

After a long, sleepless night of trying to muffle the sounds of drunk truck drivers struggling to find their hotel rooms in near darkness, I stumbled out of my room, desperate for coffee. The rain was coming down in sheets, and Mesfin was running late. My driver (and now friend), Zeyede Nigat, ordered tea. But I was resolute. I was waiting for forest coffee.

About 25 minutes later, Mesfin showed up and we headed into town. The rain had slowed to a light sprinkle, but everything was sticky with mud. My boots sucked into the ground with every step and into the earth of the small shack where we had our coffee. Despite opening onto a crowded, busy street, the little shop—a two-walled structure covered with a muddy plastic tarp—felt cozy.

We sat on the wooden bench closest to the fire, and a woman with a headscarf and a broad, gap-toothed smile greeted us. I understood nothing of the exchange except the word *bunna*: coffee. That was all I needed to hear.

While Mesfin and Zeyede chatted, I wandered to the area behind the shop and tried to casually watch the woman crouched down in the mud, tending to her *jebena*. She was welcoming, but I was embarrassed that I didn't speak her language and, eventually, returned to our wooden bench under the green tarp. The coffee arrived soon after, accompanied by a cup filled with large granules of sugar that reminded me of chai stalls in India.

The coffee was exactly as Mesfin had described: earthy and strong with a lingering finish. I wasn't sure if I liked it or not. My palate was still learning, but I could tell part of the reason the coffee tasted muddy was because of how it had been fermented and dried. Since forest coffee doesn't command the price premium of other, better known places, the returns are low. As a result, people are unwilling to invest much in its processing, which reinforces a vicious cycle of low prices because the quality can't match that of one of Ethiopia's prized Harrar or Sidamo coffees.

Despite my hesitancy, I wanted to cultivate a taste for it. By learning to love what was wild, I could—we could—increase the premium for the coffee and, in turn, improve its quality. We could, I understood better than ever, help save it by drinking it.

"You relate coffee flavor to what you know," Mesfin explained. "If a natural phenomenon is considered a defect, it's a disconnect from nature. Shouldn't flavor also reflect the coffee ecosystem? The abundance of coffee from the forest can be an indicator of healthy production. Taste should not be separate from this sustainability." He then added, "The same coffee in different hands brings a big difference in quality."

I gasped in agreement. I had come to Ethiopia to understand those hands. That's how I ended up following two farmers—Tadesse Gudina and Tebeje Neguse—to the patch of forest where they collect wild coffee.

Tadesse and Tebeje are strong, sturdy men; their hands are dry and cracked, their fingernails crusted with mud. One wore knock-off Crocs and the other cheap sneakers, but both moved through the forest with grace, climbing effortlessly over tree limbs, quickly navigating the uneven terrain as I—in professional hiking boots—lumbered behind. They laughed as they hiked up the hill and spoke passionately in Oromo, their native language, about the coffee plants they identified for harvest. Tadesse pointed out the mother coffee trees and emerging seedlings in close proximity, reminding me of what Mesfin had said when I commented that the semi-forest areas looked pristine: "Human hands are here, Simran."

As we reached the plateau—that fuzzy line between the forest and its semi-forest periphery—I asked Tadesse and Tebeje if I could take their photos. We took a few standard shots and then I moved in for a close-up:

their hands. "I want people to see the hands that harvested their coffee," I said. They looked at me like I was a total weirdo, but obliged and allowed me to take tight shots of their hands clasping branches stippled with tiny white coffee blossoms.

When I asked them what they wished people knew about their lives, the gregarious men fell silent. It had taken me nearly 12 hours to reach them, driving from Addis Ababa with Moata Raya, the technical advisor for the NGO TechnoServe Ethiopia, on some of the worst roads I have ever experienced. "Thank you" was all Tadesse and Tebeje said. "Thank you for traveling here to share our story."

Farmers are often talked *about*. They are the silent stars of exotic pictures that accompany tasting notes for specialty coffees; they are distilled into the singular portrait of Juan Valdez, the icon of bulk coffee. But rarely are they spoken *to*. I had briefly considered, every morning, how they influenced my life, but I never thought about how I influenced theirs. How my actions—and the actions of every coffee drinker— might also be part of their morning prayer.

According to Peter's rough calculations, coffee passes through at least 18 pairs of hands to get to our cup. "Farmer, picker, wet-miller, patio-drier, dry mill worker, cupper/lot maker, bagger/loader, trucker, container loader, ship driver, container unloader, warehouse worker, cupper, buyer, roaster, bagger, trucker, barista. There's 18," he wrote, "leaving out any financial professionals and probably a few more truckers."

The first sets of hands cultivate: They plant seeds, nurture saplings and wait for the bright red, orange or yellow cherry—the fruit that made Kaldi and his goats frisky. The average coffee shrub takes three to five years to bear fruit and produces its highest yields in the 10 years that follow. Chemical fertilizers and pesticides are often limited, not because of ethics or the price premium an organic certification might command but because of the limited financial resources of the small-scale farmers who raise 95 percent of the crop. Most Ethiopian farmers are too poor to buy them.

Generally speaking, "growers get about 10 percent of the final price consumers pay," explains agricultural economist Mick Wheeler. "They suffer the greatest price volatility and risk across the board. But they are also the ones who add more value to the product's worth. They deserve more of the cake."

During harvest, the work is unrelenting. Laborers rotate among trees every week to 10 days, carefully selecting fully ripened fruit. The cherries, which are slightly larger than a cranberry, are picked by hand, one by one. A solid day's work reaps a harvest of 45 to 90 kilograms (100 to 200 pounds), which ultimately yields about 9 to 18 kilograms (20 to 40 pounds) of ripened, unroasted (green) coffee. In season, migrant workers are given food and lodging and earn approximately 5 cents for every pound of coffee they pick. The crop is harvested once a year, and it takes more than 7 pounds of cherry to make 1 pound of roasted coffee.

After picking comes processing, which, again, has its roots in Ethiopia. The oldest method is the "natural" process in which berries are dried in their skins. The practice requires significant labor and attention, but doesn't use machinery or water, which is why it's still the dominant method of processing in Ethiopia. The cherries are sorted, cleaned and spread out in the sun to dry on concrete or brick patios or raised beds.

Depending on weather conditions, they are raked or turned by hand four to five times a day (and covered at night or when it rains) for anywhere from two to four weeks. This ensures even drying and discourages the development of fungi. It's a Goldilocks moment: Too little drying and the coffee seeds get moldy; too much drying and they can become brittle and break. Because the mucilage—the sticky layer of pulp surrounding the seed—has longer contact with the seed, the process tends to yield a brew with a wilder, more intense smell and flavor. But longer contact with seeds isn't always a good thing. Sometimes the taste is harsh and the aroma is funky, almost rotten. But when the taste is good, it is intensely sweet, fruity and earthy, like the concentration of flavors in a raisin versus a grape.

For a more uniform and predictable cup, the skin and mucilage are partially or fully rinsed off in what is known as a "washed" (or "wet") process. This is the kind of coffee that requires attention; the flavors are more nuanced, like the coffee you find in places like Seven Seeds.

The acidity, body and flavor are more balanced, and the taste is clean. Not surprisingly, this coffee commands a higher price because it's more labor-intensive and needs a considerable amount of water—another living-in-the-tension kind of moment: Coffee is grown in places where people don't have continuous access to water or electricity. Should their limited resources be spent on coffee processing? Is the price premium worth it? Can we direct more of our money toward shade-grown coffee, the most water-friendly way to grow the crop?[83]

In washed coffee, the two seeds within each cherry are pushed out of their skin, and any pulp that still clings to the seed is removed directly, either by washing or through fermentation (followed by washing). For all coffee, the seeds are then dried, the papery layer (parchment) is hulled off and the thin, silver skin is polished away. As with cacao, the seed has now died, transformed into what we (incorrectly but universally) call the bean.

The coffee is, again, cleaned and sorted, then graded and prepared for export or local sale. From there, it's roasted, ground, brewed and consumed. The flavor is inherently good, Moata Raya says. So the goal is to "exploit what the bean has." At every juncture, those intrinsic qualities can be enhanced or compromised.

I now understand what Aaron said about the whole story being revealed in the cup. Coffee has to go through a series of processes and changes to reveal its true identity. If we pay attention, we can taste decisions on species, variety, processing, milling, roasting and brewing. Every step—every set of hands—matters.

When I was in the forest, I asked Mesfin what the men and women who gather wild coffee were looking for, how they selected the best coffee cherries. He smiled: "Look for the mother tree." He pointed to a tall tree with a thick trunk, the kind you'd never see on a planned farm. "The people who come here are looking for trees with a lot of life around them." I bent down and touched the tiny saplings, the vibrant baby coffee. "They look ahead because they want to pass this on to the next generation. But with modernization, now people look for high production and go first for the market."

That market is guided by what economist Adam Smith called "the invisible hand"[84]—the unseen force that manages the supply and demand of goods. The hand that impacts the present and future of coffee.

Mass-consumer coffee (think big brands sold in giant tins in grocery stores) is sourced in bulk on commodities markets, unlike specialty coffee, which involves smaller volumes. Both types of coffee value quantity, but while commodity coffee is defined by these yields, premium-grade specialty coffee prioritizes specific flavor profiles. "Specialty coffee," Peter reminds us, "is about diversity in flavor. We aren't trying to get everything to taste like *one* thing. We favor multiple types and places. We celebrate it." But within that diversity, farmers and breeders still select certain varieties over others—that's the nature of agriculture. Bulk and specialty coffee result in varying degrees of genetic bottlenecks because they both select for specific traits. (This is why preserving wild places—and the crops and crop wild relatives within them—is so critical.)

Characterized by volume or weight, commodities markets require minimal variation from producer to producer. In economics terms, they're "fungible," able to replace or be replaced by another identical product. Within this context, coffee is a financial instrument—a thing—not a dynamic, living plant.

While some traders know a lot about coffee (or whatever crop they work with) and have a commitment to setting a fair price, they are far removed from the people who touch it—farmers, intermediaries, exporters, coffee roasters and buyers. They know coffee as a discrete good. They sit far away from parched earth, diseased plants and verdant forest, yet they have tremendous influence over what farmers and those involved in producing coffee get paid for their efforts—and how much we ultimately pay for our cup.

Bulk coffee is bought and sold in what's called the soft commodities market, alongside cocoa, cotton, wheat, corn, sugar and other products that are grown instead of mined. But this market isn't a place in the traditional sense of the word; it's a virtual network.

Traders, at one time, gathered on the floor of their respective exchanges and made transactions face-to-face. This is still done in local exchanges, such as the Ethiopian Coffee Exchange (ECX). Buyers and sellers gather in a large, glassed-in room in Addis Ababa and confirm trades with the touch of hands (what, to me, looked like an ongoing sequence of high fives). Global commodities traders, on the other hand, sit in front of computer screens and buy and sell coffee as one of many transactions in a day.

Coffee and all other soft commodities (including cocoa) are traded in two basic ways. The first is to use the spot market for immediate delivery. The price, paid on the spot, is a reference point for the current value of beans, based on that day's supply and demand.

The second method of transaction is through the futures market, trading commodities that don't actually exist but will in the future. The two most heavily traded exchanges are in London and New York City: The London International Financial Futures and Options Exchange (LIFFE) sets the benchmark for robusta coffee, while the Intercontinental Exchange (ICE) in New York City sets the price for arabica. Beans from specific locations of certain grades are bought or sold for delivery at a designated point in the future, when, legally, they have to be physically available.

Or perhaps it's safer to say they might be available in the future. These contracts are bought and sold many, many times. The challenge is, coffee that *doesn't yet exist* influences the price of coffee that *does*. This is because the current price of coffee is informed by the total number of contracts traded (including futures), "which far exceeds the physical amount of coffee that changes hands."[85]

Dizzying, I know. But consider the implications. For coffee *consumers*, this impacts our future coffee choices as much as climate change, pests or disease. For coffee *producers*, this determines the kind of coffee they grow (whatever grows fastest and most abundantly) and how much it will earn them, which, again, impacts what we drink now and in the future.

The upside of the futures market is that it's a way for commodities producers to manage the risks associated with changes in coffee prices by locking them down in advance. A producer—not a small farmer, but a larger producer or cooperative that has the resources to access this market—can leverage futures to lock in a price today for a crop that ripens a few months down the road. Conversely, a coffee buyer can set a price today for his or her future bean needs.

There are two critical things to remember about coffee markets. The first is that multiple factors influence the price of coffee. Coffee is subject to diseases, pests, weather, political tensions, transportation costs, and local and global economic activities.[86] The second is that the farmers who grow the crop are largely overlooked by those doing the trading.

One coffee trader I spoke with reminded me that the intermediaries between traders and farmers (like the middlemen buying and selling cacao) also require scrutiny: "When prices are high, these distributors fail to pass on to small producers the full price they are getting in the marketplace. When prices are falling, they focus more on maintaining their market share [by lowering prices further] than they do on making sure the coffee price doesn't fall below the cost of production." But he cautioned against demonizing the entire system, reminding me that "the commodity markets are just the *mechanism* through which these activities are played out." Behind the mechanisms are people.

At one time, farmers were protected by the International Coffee Agreement (ICA). "The aim was to keep the price of coffee relatively high and relatively stable, within a price band or 'corset' ranging from $1.20/lb. to $1.40/lb. To prevent oversupply, countries had to agree to not exceed their 'fair' share of coffee exports."[87] But if prices increased to more than $1.40/lb., "producers were permitted to exceed their quotas to meet the surge in demand."[88] These export quotas were indefinitely suspended after 1989 when the International Coffee Organization (ICO) failed, multiple times, to agree on terms during negotiations.[89]

Peter Giuliano put this in perspective in the documentary *A Film About Coffee:* "In 2000 ... the bottom fell out of the coffee market. ... Coffee went from being at a normal level of, say, $1.25 a pound for green to 60, 50 cents a pound. And that was way below the cost of production for coffee farmers. Suddenly, we were looking at actual famine in the coffee world."[90] By 2001, prices sank again, "totaling less than one-third of their 1960 levels." This precipitous drop affected over 25 million families and crippled the economies of coffee-producing countries.[91]

But one of the perversions of economics is that *higher* prices don't necessarily improve the lives of farmers or reflect anything worth cele-brating, either. In 2014, arabica coffee futures skyrocketed 50 percent, "as a drought in top grower Brazil propelled prices of the beans to their first yearly gain in four years."[92] It was the worst drought the country had experienced in decades.

I wondered why I hadn't known about these dramatic shifts—and why the price of my latte had held constant. It's because these shocks are absorbed by those with the least amount of leverage and power: small farmers.

Markets are full of monopolies, and coffee is no exception. Five corporations—Kraft General Foods, Nestlé, Proctor & Gamble, Sara Lee and Tchibo—control 50 percent of the global coffee market: one out of every two cups of coffee consumed.[93] These corporations, which own Maxwell House, Folgers, Chock Full o' Nuts and Hills Brothers, push prices down through what's known as economies of scale, buying massive quantities of coffee for less money. Small-scale coffee buyers can't compete.

This monopoly also extends to coffee producers. In one decade, Vietnam increased its production of coffee by 1,400 percent, flooding the market with cheap robusta to become the second-largest producer of coffee in the world.[94] Meanwhile, Brazil, the world's largest pro-ducer, alternates between high- and low-output years, yet continues to influence prices in countries that have consistent annual output. That leads to the most perverse part of the economic equation: Farmers are rewarded for increased productivity with lower prices.

Supporters of globalization say the market will sort out these problems, but the market *caused* them. The commodification of coffee pushes farmers to grow as much as possible by whatever means possible. This has resulted in deforestation, habitat loss for plants and animals, and the slow but steady loss of agrobiodiversity. We tell farmers the way to make more money is to grow more and sell more—at any social and environmental cost. As a result, most have no choice but to opt for growing methods and coffee varieties that increase volumes *now*: high-yielding varieties grown in full sun. And sadly, agricultural economist Mick Wheeler explains, "there's no safety net for small producers; they don't get to say no."

According to the Fairtrade Foundation, "this price volatility has significant consequences for those who depend on coffee for their livelihood, making it difficult for growers to predict their income for the coming season and budget for their household and farming needs. When prices are low, farmers have neither the incentive nor resources to invest in good maintenance of their farms by applying fertilisers and pesticides or replacing old trees. When prices fall below the costs of production, farmers struggle to put adequate food on the table and pay medical bills and school fees—a major reason for children being taken out of school to contribute to the family income."[95]

Ethiopian farmers—stewards of a global crop that, according to the ICO, brings in more than $174 billion annually—earn less than 19 cents per pound of ripe coffee cherry. When the market slumps, farmers earn *less* than the cost of production. Not only are they working for nothing, they're essentially paying to grow the crop.

Studies assert more than 99 percent of the biodiversity of arabica coffee is contained within Ethiopia's and South Sudan's forests. While this percentage is currently being reevaluated (emerging research shows it might be less than expected[96]), what *is* certain is that no one is being compensated for preserving this biodiversity.

"Then why do they keep working in coffee?" I asked Mick.

"Because they have no other choice," he said. "Coffee is the perfect smallholder crop. Agronomically speaking, it's easy. You don't need an advanced degree to grow it. And once you pick it, you can dry it as a cherry or process it further. Then you can store it and wait for a buyer to come around."

Within Ethiopia, the places that are good for growing coffee are also ideal for growing another stimulant: khat. Khat is a large shrub native to the Horn of Africa and Arabian Peninsula whose leaves can be brewed into a tea or rolled up and chewed over the course of hours. It's a stimulant that tastes leafy and acts a lot like speed. Khat is harvested every two to four weeks, and one pound commands a minimum price of 50 birr, or $2.60 (versus the 19 cents per pound farmers receive for coffee). This punishing equation is why farmers hold back a portion of their harvest to ensure they have coffee in the off-season. "Coffee," Moata reminds me, "also staves off hunger."

This makes my heart ache. I have the power to choose when and what to drink (and eat). I open up my cupboard and, on any given morning, select from a small array of coffees from Ethiopia, Kenya and Rwanda— and that's just from the continent of Africa. My choice to reach for one over the other has real, significant consequences.

Some would argue caring for farmers isn't the responsibility of the market and that it's unfair to blame retailers and traders for the decisions farmers make to increase productivity or sustain themselves. But then what *is* our responsibility to take care of one another? These farmers' efforts help me enter every new day. Their work boosts *my* productivity. For pennies on the dollar, they preserve the genetic diversity required to ensure I will be able to drink delicious coffee for the rest of my life. What have I done to support them? Not support in that generic "I pay a price premium" sort of way. What have I done to encourage the farmers of Ethiopia to grow coffee, not khat?

What is my responsibility to them—to the women and men, to the land that is directly responsible for the past, present and future of coffee? How do I acknowledge their value in tangible ways that improve their lives? And what should a cup of coffee cost? What is its real value? If the answer is nothing more than what we currently pay, it's time to reach for a Red Bull.

I desperately want a simple solution, a less complicated morning prayer. But there isn't one. We have to grow comfortable with many answers, not one. And to remember what Peter said: "Delicious coffee that hurts the land and people isn't delicious." The power is in our hands, in our morning coffee, and in every sip that follows.

This power, it's audacious. It can change everything.

Phyllis Johnson, the former vice president of the International Women's Coffee Alliance, told me that many coffee buyers she met "wanted to know the price, not the people." Her work has taken her all over the world to meet some of the industry's poorest subsistence farmers. "It's a little depressing somehow to sit in an American café when you understand what's happening. We walk into the coffee shop complaining about simple things when the people growing the coffee don't even have their basic needs met. Sometimes I want to say, 'If you only knew.'"

The place that offers greater opportunity for us to properly compensate farmers for their hard work is the cup that costs closer to $3.99, not 99 cents. And not just any $3.99 cup of coffee; they're not all the same.

The model of compensation and certification program, known as fair trade, guarantees coffee growers a set price prior to harvest, rather than one that's decided after they've already grown and harvested the crop (what essentially boils down to clocking in to a job every day but having no idea what you'll get paid). It is a critical first step.

But while this program has created a solid price floor and expanded small farmer cooperatives, critics argue certification is costly and labor-intensive, and it doesn't address ecologically friendly cultivation methods (including growing in shade or preserving rainforests) or reward quality. Although environment challenges and taste weren't what fair trade set out to do, they are critical parts of the coffee continuum: We want to support, sustain—and drink—something delicious. The latter is one of the most compelling reasons behind the development of what's known as direct trade.

Direct trade isn't a codified system but, rather, a grassroots model that's been developed in specialty coffee through direct relationships smaller buyers have built with smaller growers. That smallness—the intimacy of a direct relationship—is important. It allows for greater accountability on both ends of the coffee value chain and is a way for coffee buyers to reward the care and diverse flavors that define specialty coffee.

The fact that direct trade doesn't have an official certification protocol is both good and bad. While it cuts down on bureaucracy and costly administrative fees, it also functions without oversight. There isn't a third party monitoring these activities, so the interaction is basically whatever coffee purveyors like Seven Seeds tell us it is. This trust, Peter says, "can be easily abused." When we purchase from places committed to direct relationships, our extra 50 cents might buy food, medicine or equipment for the farmers. It might help dig a well, build a bridge or strengthen a relationship. Or it might not. We have to ask these questions of the people who provide our coffee. These relationships belong to us, too.

The way to stem the loss of coffee is to recognize its value—and pay for it. It's hard to remember this when we're trying to get ourselves awake enough to get to work or already blanching over the cost of a pour-over. But the truth is, we're not paying enough. This is why khat is pushing out coffee in Ethiopia, in Yemen and in other coffee-growing countries. Whatever we're paying isn't enough for a crop whose price is largely determined on trading floors or in front of computer terminals where traders buy low and sell high. The richness of soil, land, culture and people isn't counted; it is erased. Coffee and every other commodity crop—and the people who grow them—are fungible.

The farmers protecting the diversity of forest coffee have not been compensated for their efforts because economic valuations of coffee resources never made it off the page of calculations. Their stewardship, as one commodities trader told me, "is economically not relevant." Maybe not according to the balance sheet, but, in life, this conservation is priceless. "What we find in the wild would take 40 to 50 years to create in a lab or an arboretum," Moata explains. "Shouldn't the people who benefit from this contribute to the area it comes from?"

I thought back to the one-hotel town, to my feet sticking in the mud, to the plastic tarp and the fake Crocs. I thought of the 18 sets of hands that bring us our coffee, the people who—in spite of poor roads, sporadic electricity and limited water—manage to bring us something magnificent.

Of course.

We save the coffee and the forest (plus the pollinators, plants and animals that thrive within it) with *our* hands, by reaching for every one of those tastes Peter said can be found in Ethiopia, expanding beyond popular coffee varieties to take a risk on what is lesser known—and on what is wild. "It's got to be all about the coffee," Aaron says. "It's not about your business model, or how fancy your packaging is, just let your coffee be coffee. I'm happy if it comes in a brown paper bag, it's just got to taste amazing."[97]

I ended my trip to Ethiopia with the farmers who made me fall in love with the country, the ones responsible for the Yirgacheffe flowers that had bloomed in my cup back in Melbourne. The journey was long and dusty but grew increasingly lush as we approached the Yirgacheffe Farmers Cooperative Union (YFCU). I was beside myself; this visit was the one I had hoped for more than any other—one that hung in the balance until an hour before the long journey back to Addis Ababa.

My driver, Alazar Mengstu, and I headed up a long, rocky path to a small room that served as the cooperative's local office. Standing inside were eight men and one woman, ranging in ages from 26 to 52, plus Andenet Bekele, YFCU's certification and quality control officer. The room felt cramped so I suggested, through Alazar, that we move outside. We set up a circle of chairs underneath a shade tree, and I began to ask them about their lives.

My main question was straightforward: How do you make this coffee taste so good? Bereket Beyene, a 26-year-old farmer in a crisp white shirt smiled and said, "God gave this coffee to this area and to the people. This elevation is blessed; it is simply here." He compared it to Brazil, the world's largest coffee producer, explaining they have "a huge number of coffee harvests but only when they work hard to make it grow." Everyone nodded in agreement; coffee is Ethiopia's birthright.

But then Alemu Seda, a 52-year-old man with a goatee and well-worn hat, interjected that, while Yirgacheffe didn't have a problem with the existence of coffee, they did face sizable challenges. "Our machines are old and broken. We have one machine from the 1980s and we couldn't find the [replacement] part, so we couldn't get the harvest. Farmers that have modern equipment conquer the market. We don't have a problem with coffee but with machines and roads. Otherwise we could deliver much more." Andenet added, "The road is broken. It's a big problem to just get the coffee to the processing center because of it."

I thought back to what Phyllis had shared with me: If we only knew. And then I thought what an absolute miracle it is that this coffee tastes so good. Coffee from Yirgacheffe is as precious to me as Scott's Trousseau Gris—delicate, almost tea-like, but still strong. It tastes like flowers but is also earthy. It reached for me—and I keep reaching back.

The farmers and I walked down to the processing area, where I saw the broken machine Alemu had described. Weeds were growing all around it. The moment was heavy but also hopeful: "No one mentioned Yirgacheffe before," Alemu said brightly. "Our time is coming now. And maybe our grandsons' sons will have a better life because of this. Thanks to God."

Before I left, Bereket added, "We are lucky the premium for Yirgacheffe coffee has increased because we need clinics, bridges, roads and schools. The world should take care of the stewards of coffee, no?"

"Yes," I responded, "we should."

Mick Wheeler retired in 2011 from his position as the executive director of the Specialty Coffee Association of Europe. He is now the overseas representative for the Papua New Guinea Coffee Industry Corporation and the country's permanent representative to the International Coffee Organization. As his resume shows, Mick's commitment to farmers is unwavering. But he admits there is a lot of work to do. "It's very sad for me to say this, but coffee producers would be better to expand production rather than wait for better prices from very small lots. Farmers have limited resources to improve quality or increase output, and unfortunately, quality doesn't pay growers to produce. Unless there is a reward for it, how can we ask these farmers to push quality? They need long-term contracts and better prices."

Just as I steeled myself around that solution, he added, "I'm prepared to say that's just not going to happen. The day farmers realize this truth is the day we will all suffer." I don't want to contemplate that day, a life without what those hands bring forth, a morning without my profane prayer.

The places where coffee comes from are poor, but not impoverished. Their stories are there in the cup, just as Aaron said, but these narratives are ultimately shaped and defined by us, the consumers.

The decision is in our hands.

III. TASTING COFFEE

Few people love coffee the first time they taste it. We develop an affinity for the drug effect of caffeine, however, forcing ourselves to drink it: Our brain links the flavors to the effects, and we begin to love it, seeing the kaleidoscope of flavors that exist behind the bitterness. We call this an acquired taste.

—PETER GIULIANO

Let your experience of coffee be fluid. Allow Peter's words to be your guide and open yourself up to flavors that are a bit more subtle. Try to start with inner and outer quietude. As coffee quality expert Silvio Leite said during one of the world's premier coffee competitions, "Please, silence during the cupping. This is about communication between you and the coffee."[98] This is your chance to hear—and listen.

Preparing for a coffee tasting is a bit more involved than wine or chocolate, but not difficult. One of the critical steps is to buy whole beans and grind them just before sampling. Small hand grinders cost less than $30 and will open up a world of aromas (one that dissipates quickly, which is why it's a good idea to grind beans close to drinking).

Curate your curiosity. Coffee reflects terroir, post-harvest processes and the skills of those who roast it. Start off trying different regions from a single roaster—or a single place expressed by different roasters. Over time, this will help you develop a baseline for what you love.

Professional cuppers grind beans to the texture of coarse sand, brew them in hot water (around 200 degrees Fahrenheit), skim the grinds off by hand, and then sample the coffee without the distraction of milk or sugar.[99] This tea-like preparation is the easiest way to find everything in coffee, including defects.

Take a look at the Counter Culture Coffee defect wheel found on page 347 and see what, if any, terms resonate. Ideally, your coffee should be absent of these qualities.

Although the Counter Culture Coffee Taster's Flavor Wheel isn't the industry standard (that designation belongs to the SCAA Flavor Wheel, which was created by the broader coffee community), it beautifully illustrates the range of what's possible—of positive smells and tastes that will help you identify not only what you love, but what you don't.

The wheel was largely the brainchild of Counter Culture Coffee's buyer and quality manager Timothy Hill. He explained in *Barista Magazine:* "The wheel is design[ed] to be very intuitive and [responsive] ... as the frequency of descriptions is based on what is popular in coffee at that moment. For instance right now ... washed Ethiopian coffees are extremely popular and gaining market share among the best roasters in the country. As those coffees tend to be very floral, citric, and fruited, this will likely mean those categories continue to dominate the wheel and potentially even expand. Coffees that are earthy and tobaccoy are losing their popularity ... so likewise those categories will continue to shrink in size and scope."[100] This is also why the SCAA and World Coffee Research are developing a lexicon (language) that will be an objective point of reference for coffee flavors (similar to Katie Gilmer's tasting exercise at UOPROCAE).

The scoring sheet Seven Seeds uses for public cuppings (found on page 348) is, for many people, one of their first experiences deeply exploring coffee; it gives a starting vocabulary for tastes and aromas in the same way flavor wheels do. Although we generally want to steer clear of good/bad polarities, the form helps quantify what feels like a wholly subjective experience.

Pair the wheel with the form and you're ready to find your words and flavors. The form breaks the sensory experience and identifications down into light, medium or heavy across four categories: 1) aroma, 2) sweetness, 3) acidity and 4) mouthfeel. "Sweet coffee," Aaron Wood explains, "is coffee that has matured well." Body, the mouthfeel of the coffee, is similar to the sensation of weight and texture found in wine, or skim versus whole milk. "We want something creamy and buttery—full," he says. Acidity is the tanginess that enlivens your mouth, the perfect counterbalance to sweet.

Before grinding the beans, look at them. Are they oily or matte? Lighter brown or close to black (the darker the bean, the darker the roast)? Imagine the long journey they've taken.

Funnel them into your grinder and listen as they are pulverized. The smells will now start to emerge. Once you've finished grinding, put the water on to boil. Then return your attention to the dry grounds. Open your mouth slightly and draw in the volatile aromas so they can go up the nasal passage, hit the olfactory bulb, and become discernible smells. Take quick, not lengthy, sniffs (bloodhound, not yogi). Write down your observations.

Once the water is nearly boiling, pour it evenly over the grounds. If you pay close attention, you'll hear a faint popping as the water saturates the coffee. (That's the sound of carbon dioxide being released, a sure sign the coffee is fresh.)

Take in that wet aroma. Then let the coffee steep for four minutes (until it is closer to body temperature), stir and, again, inhale. "The coffee grounds," Peter clarifies, "having been liberated from their carbon dioxide bubble balloons, will sink to the bottom of the cup." Lift any remaining grounds or foam out with a spoon.

Smell, listen, look: In finely ground espresso, light foam (known as crema) is considered a mark of quality. There is usually a lot less of it in coarsely ground coffee, but a bit of foam might be in your cup.

And now taste, slurping a spoonful to increase aeration. Let it roll around on your tongue. Notice how the sweetness and acidity manifest. Are they balanced? Does one dominate? See what rises—and what falls. Then, feel free to spit out the brew: Hyper-caffeination is at odds with coffee contemplation. Before resetting your palate with a sip of warm water, notice what's lingered and what's there at the finish.

Be prepared, at first, for coffee to simply taste like coffee. Be patient and compassionate with yourself and the brew. A good place to start is with natural coffees from Ethiopia. "They can be funky or fruity and have a distinct blueberry flavor," Peter explains. "They aren't nuanced; they're intense and make for an easy connection."

Take another sip. Close your eyes. See what's revealed. Is the flavor sharp or soft? Does it make your mouth feel dry or cause it to water? Does the flavor linger? If so, do you welcome what stayed or wish it had disappeared? Is the finish clean or dirty? What story is emerging in your cup?

Tasting is different from drinking. It's deliberate and focused and will help build sensory connections so you can find them more easily in your daily cups of coffee. These flavors will also change depending on how coffee is brewed and served. They are dynamic, like the coffee seeds from which they came.

Cup and contemplate.

BEER

I. THE SPICE

"Respect the Beer."

This is what the small lavender Post-it note above my writing desk says. I placed it there as both proclamation and inspiration because I hated beer. I tried, many times, to love it—but I didn't. My friends were incredulous when we'd visit breweries and I nursed a sampler while they downed a pint. They couldn't understand why I sipped beer like whiskey while they gulped it like water.

I didn't, either. It might have been a sensory flashback to my first sips of Bud Light in junior high school. Or the bitterness of hops that overwhelmed me whenever I tipped the small glass to my mouth. I couldn't explain it—but I wanted to understand it.

"When people say 'I don't like beer,'" Garrett Oliver, acclaimed author and brewmaster at Brooklyn Brewery in New York, says, "that's usually a sign that they have only tasted one type of beer. It's like hearing a single song and saying 'I don't like music.'"[1]

Beer is a remarkable beverage: a liquid as old as human civilization made of four simple, relatively inexpensive ingredients; one that can be made with any grain, in any place. Andean communities, for instance, make beer from corn, root vegetables and fruits, while Japanese make sake—mistakenly identified as wine rather than beer—from fermented rice. Beer doesn't belong to a single culture or geographical area. It's democratic and belongs to everyone.

That's why, despite my disinclination, I wanted to learn more.

Even if I didn't love it, I wanted to take in every lesson the tiny beer I nursed had to teach me and understand the decisions that culminated in the glass.

Beer is composed of water, plants and a microbe, plus additional ingredients used for flavoring. The plants include a grain from the Poaceae family, one of the largest botanical families in existence. Barley is most common, but brewers also use wheat, rye or corn. The other plant featured in beer is hops, part of the hemp family known as Cannabaceae, which also includes cannabis.[2]

When aficionados boast of beer being both art and science, they aren't kidding. Every batch is an experiment in microbiology, botany, chemistry and physics—beginning with microbes in the water and ending with the surface tension of the glass into which the beer is poured. A lesson in physiology kicks in when we take in the beer through our senses.

I didn't expect this much science. I was born in Bavaria in southeastern Germany—the birthplace of Oktoberfest and beer gardens, where beer costs less than water—to a father who loves a good Hefeweizen and a mother who was advised to drink the occasional beer to rid herself of insomnia when she was pregnant with me. Despite my in utero exposure, beer wasn't part of my life. The bulk of my experience was limited to the kegs my friends and I tried to finish in high school. I didn't realize beer was so ... heady.

It is: "Beer may be humble," Randy Mosher writes, "but it is not simple."[3] Randy is the author of *Tasting Beer: An Insider's Guide to the World's Greatest Drink*, a book I carried around with me from place to place like some kind of beer bible. Unbeknownst to Randy, he became a constant companion on my journey.

With Randy's words echoing in my head, I enrolled in Beer School at Foothills Brewing, the oldest and largest microbrewery in Winston-Salem, North Carolina. On the first Saturday of every month, Foothills offers an immersive lesson in beer—from history to brewing to tasting. As I listened to instructor (and Foothills' chief operating officer) Sarah Bartholomaus describe the beers we sampled—pilsners, porters, stouts and ales—I kept asking myself, "What's not to love? Why don't

I like beer?" And then we sampled the Foothills Bourbon Barrel Aged Stout—a beer that tasted faintly of bourbon and offered a sweetness that tempered the bitter. It gave me a glimmer of hope; even *I* could grow to love this beverage.

But, really, it doesn't matter. It doesn't matter if I like beer, and it doesn't matter if you like beer. To understand it is to appreciate it—not just beer, but every food and drink. This appreciation is essential. It's the most important lesson I learned from every taste expert I met: Taste *everything*. Let the experiences help clarify your likes and dislikes, sure, but also let them inform the places in between. Let them teach you more about yourself.

My goal was to respect the beer—and stay open to the possibility of, some day, loving it. But before I could love it, I had to know it. I started my journey at the National Collection of Yeast Cultures (NCYC) in Norwich, England, on a typical cold and grey English morning in early spring 2014, to understand how one microscopic organism transforms the beverage I was trying to know and love.

I arrived nearly 30 minutes early. Unlike my overpreparation for the wine interview with Ann Noble, I felt painfully underprepared for this one. I had just one partial page of questions, and it was handwritten, not typed. The receptionist suggested I head to the cafeteria for a cup of tea and told me she'd have the team meet me there. Team? I was expecting to interview one person: collections manager Chris Bond. I made myself a cup of Earl Grey, took a seat in front of one of the many silkscreens of magnified yeast cells that adorned the room, and started reviewing notes on yeast as if I were cramming for an exam.

In beer, yeast is everything; it's the catalyst that turns a soupy sludge of grain and malt into the alcoholic nectar 41 percent of Americans imbibe regularly.[4] Yeast impacts not only beer but bread, wine, coffee, chocolate and every other life-form on the planet. It is conserved in places such as the NCYC, one of the largest freestanding *ex situ* collections in the world.

Exactly two minutes after I sat down, a tall, affable man approached me and introduced himself as Steve James, a molecular taxonomist and deputy curator of the collection. "Why don't you join us at our table?" he asked. So much for cramming.

I packed up my notes and joined Steve and his colleagues, including two microbiologists, a biofuels researcher and Chris Bond—a bespectacled, serious-looking man with long nails and sharp wit. Within a few minutes of meeting Chris, I was already referring to him as a rock star. He was *my* rock star, another person dedicated to saving what was precious, a custodian of the substance that made beer beer and bread bread. (I later learned Chris is *literally* a rock star, a member of the electronica band Army of Mice.)

Chris is responsible for the intake of yeast at the NCYC, plus its duplication and cataloging. He also oversees quality control to ensure the purity of the collected strains. "When yeasts first arrive," he said with a dark smile, "they might have a small amount of bacteria in them. If I see bacteria, I kill it. This is a yeast collection, *not* a collection of bacteria." I love this man.

"We have yeast from pole to pole," Steve added. "The clouds hold bacteria, and the deserts and sea contain yeast."

Chris deadpanned: "They seem to have colonized all areas."

The banter between Steve and Chris was so affectionate and funny, they dominated my interview notes. Halfway through the conversation, I started to refer to them as simply YB: the Yeastie Boys.

"Have you ever thought what would happen if you lost all your yeast?" Chris asked. Never. I had never in my life considered such a question.

But here's why it's so interesting. Microbes, including yeast, are *everything*. Not just in beer, but in life. The study of yeast is the study of us. Cosmologist Carl Sagan once wrote, "The nitrogen in our DNA, the calcium in our teeth, the iron in our blood, the carbon in our apple pies were made in the interiors of collapsing stars. We are made of starstuff."[5]

Beautiful—and true. But we evolved from and are primarily made of microbes, also known as microorganisms—tiny (micro) living creatures (organisms) invisible to the naked eye that make up the largest number of organisms on the planet. Of the six classifications of kingdoms, half are dedicated to microscopic life-forms, while the other half are divided between plants, animals and fungi (which includes more microorganisms). Although our bodies contain about 100 trillion cells, only 10 percent of them are human. The rest are microbes, weighing in at about 3 pounds in the average adult (about the same weight as our brain).[6]

From the moment we pass through the birth canal and take our first suck of breast milk, these tiny creatures are hard at work seeding our gut and building our immune system.[7] They are the first—and most critical—link in our personal food chain, helping us digest and extract nutrients from food and protecting us from harmful microorganisms that can make us sick.

After everyone finished their tea, we headed up to the conference room. I hadn't had time to settle in and was anxious that I had only a handful of questions for the group of experts before me. I excused myself and headed to the bathroom. The last thing I expected to find there was a flier on chlamydia prevention. Chlamydia is sexually transmitted *bacteria*, not yeast. I had my in. I walked out of the bathroom, past the life-size model of a DNA helix and back to the conference room. I took my seat and launched in: "Choose chlamydia screening?"

Steve laughed and said, "Is that what you're going to write about in your book?"

"Of course," I answered. That flier set the stage for a three-hour conversation punctuated by George Clooney, rock and roll, and beer.

First, Clooney. To underscore how ubiquitous yeast is, Steve put together a PowerPoint presentation he shows in schools. It includes a close-up of a bee, a frothy mug of beer, a cluster of grapes, a horse and a headshot of George Clooney—just a few of the places where yeast appears.

Yeasts are from the fungi kingdom, part of the classification of microbes that includes bacteria and algae. "[They] are found in every environment," explains the American Academy of Microbiology in their report *If the Yeast Ain't Happy, Ain't Nobody Happy*. "On grape skins and human skin, in the stomachs of bees and in tree sap, from the bottom of the ocean to the soil under our feet."[8] Yeast is in the earth, air and water; on plants, animals and (for a period of time) inanimate objects; in and on our bodies—and George Clooney's. They fortify our immune system, support digestive health and protect our cells against oxidative stress.[9] We can't live without them.

Alcohol can't live without them, either. In beer and wine, yeast is responsible for converting sugars to alcohol and creating the carbon dioxide that results in beer's effervescence and foam. As Sarah, my beer professor at Foothills, explained, "Yeast are eating, drinking and being merry ... and putting off two by-products: alcohol and CO_2." Until yeast enters the picture, beer is just a broth of grain and water called wort (pronounced "wert"). It's yeast—and the magic of fermentation—that turns wort into fizzy alcohol. While hops are added to balance out sweetness and heighten the aroma, in some beer styles, yeast accounts for up to 70 percent of overall flavor. Without it, we'd all be drinking flat near-beer.

"The problem is, we have only the vaguest sense of [the microbes'] agenda," explains food writer Rowan Jacobsen. "Despite their omnipresence, microbes are hard to study. They are inconveniently small. The traditional solution has been to isolate a single species and culture it in a lab. But microbes, just like other forms of life, live in communities." He continues, "What's really interesting about any organism is how it responds to other organisms—does it eat them, run away from them, or start a business with them?"[10]

In beer, yeasts go into business with grains. "Yeast cells are fantastic little chemical factories," writes my virtual beer guide Randy Mosher. "They must find food, metabolize it into energy, synthesize proteins and many other molecules necessary for life, rid themselves of waste, and create more yeast. Think of them as little sacks of goo with membranes porous enough to allow some molecules through but not others."[11]

Historically, people didn't understand these "sacks of goo." They knew *something* was responsible for turning the sweet, grainy precursor to beer into alcohol, but they weren't exactly sure what it was or how it happened. Early brewers chalked the process up to magic or divine intervention, summarizing both the mysterious substance and sacred transformation with the term "Goddisgoode." (It explains why many have revised a statement Ben Franklin made about wine to "Beer is proof that God loves us and wants us to be happy."[12]) In order to make beer, people used to backslop the brew—transferring "Goddisgoode" from one fermentation into subsequent batches—hoping God would take care and make a good drink.[13]

In 1860, French chemist Louis Pasteur discovered specific organisms were involved in fermentation, defying conventional wisdom that fermentation was the by-product of the life and death of cells. He, instead, directly linked fermentation to a live organism—yeast. Two decades later, fungus specialist Emil Christian Hansen, working in the Carlsberg brewery in Denmark, found these yeasts were not one, but many, kinds of fungi.[14] By isolating a single strain, combining it with sugars and growing more in his laboratory, Hansen produced a pure culture and revolutionized how brewers made beer. He removed the unpredictability of wild yeast and turned the mystery of fermentation into a replicable, scientific process—one brewers could rely on with or without godly intervention.

Saccharomyces, yeast's scientific classification, literally means "sugar fungus." The yeast Hansen isolated was called *Saccharomyces carlsbergensis* (but was later reclassified as *Saccharomyces pastorianus* in honor of Louis Pasteur) and is still used in beers today.[15]

These fermentation yeasts were some of the first microbes ever identified, isolated and cultivated. In 1966, *Saccharomyces cerevisiae*—the species of yeast used for brewing and baking—was the first organism with a nucleus to have its genome completely sequenced. Genome sequencing is basically a way of unlocking all the secrets of an organism. The reason yeasts were sequenced first is because understanding their secrets enables us to unlock our own.

Although yeasts are single-celled organisms, their cellular structure and function are similar to human cells and those of other complex organisms. Their cells divide in the same ways ours do, except at a much faster speed. And 20 percent of our disease genes have yeast counterparts,[16] which is why this is the microbe we turn to in order to understand how to heal ourselves.

Before I learned this, I didn't know why *Saccharomyces cerevisiae* was so important. I'm not a brewer or a baker. The only yeast I had ever thought about was from a completely different genus, *Candida albicans,* the one responsible for yeast infections and not something I felt particularly grateful for. But yeast is something for which we should have immense gratitude. These microbes have helped shed light on a number of basic life processes, and are used extensively in pharmaceutical research and to help scientists understand neurodegenerative diseases, like Alzheimer's and Parkinson's. This isn't just because of some similarities to humans but because, like all microbes, they multiply like crazy, creating vast amounts of material to study and analyze.

Yeasts typically reproduce asexually through cell division, by what's known as budding. When a yeast cell is fully grown, a small swelling, or bud, forms on its surface. These buds break away from their parent cells to form new, fully formed daughter cells and, in about 20 minutes, can create an entirely new organism. However, under stressed conditions, a yeast cell turns to sex (don't we all?) and fuses with another cell.

Yeast divides at a rate of once every two hours, while human cells divide once every 12 hours. That's why, in yeast terms, aging is defined by the number of times a cell divides, rather than by chronological timing. "Unlike their mothers, the daughters start from scratch, having the potential for a full life span. Thus, individual cells are mortal, while the yeast population is immortal."[17]

I love this. Although microbes have very short life cycles and are highly susceptible to change, they stubbornly endure. Yet, because they are constantly evolving and adapting to changes in the environment, the need for stable, reliable sources of yeast is great—and is why an *ex situ* collection like the National Collection of Yeast Cultures is vital.

Founded in 1951 as a backup collection for brewers, the NCYC now houses over 4,000 different strains of yeast, belonging to more than 530 species used for brewing, baking, medical research, food science and biofuel production. In addition to maintaining the collection of yeast, the NCYC—which is partially funded by the government of the United Kingdom—identifies and sells yeast, runs a biorefinery and stores transgenic (genetically engineered) yeasts that are privately deposited and under patent.

The strains that are accessed frequently are freeze-dried and preserved in tiny glass ampules on racks placed in a climate-controlled room set at about 3.9 degrees Celsius (39 degrees Fahrenheit). An additional 12 samples are sealed in plastic straws stored in liquid nitrogen. They are in, what Chris calls, a state of "suspended animation" where no further changes can take place.

The pure yeast samples are known as strains, while the blends are known as cultures. The Collection represents about 25 to 30 percent of all the world's fully identified yeast, which is roughly 1 percent of all the different types of yeast in existence.

"Microbial communities are so complex—there may be a thousand different species in your gut or in any cubic inch of soil—that it's virtually impossible to tell who's doing what to whom," writes Rowan Jacobsen in "The Most Delicious Lab." "Which of the 99 different species in the vicinity caused that mold to start producing antibiotics? What made this dormant virus start reproducing like mad?"[18] This is, largely, a mystery.

A 2011 study suggests our planet is home to 8.7 million species, but only 15 percent of them have been fully classified. When it comes to fungi, we have identified only 7 percent.[19] This matters, because we're facing the highest species extinction rates we have ever seen. "It's frightening but true," confirms the Center for Biological Diversity. "Our planet is now in the midst of its sixth mass extinction of plants and animals—the sixth wave of extinctions in the past half-billion years. We're currently experiencing the worst spate of species die-offs since the loss of the dinosaurs 65 million years ago."[20]

And these losses are among what we *know*. Many organisms will disappear before we can ever understand or document them. This includes our closest biological relatives: primates. Roughly 90 percent of these animals—including chimpanzees, with whom we share 95 to 98 percent of our DNA—live in tropical areas that are home to at least 80 percent of the world's remaining biodiversity and are under constant threat of deforestation.[21]

A 2008 study from the International Union for Conservation of Nature found almost 50 percent of these primates are in danger of extinction.[22] This loss guts me; it makes my throat tighten and my heart and bones ache. This is what we're doing to our closest genetic allies. And what we do to them, we do to ourselves.

Relationships in nature are symbiotic. Fruit flies, for example, are attracted to a smell emitted by certain types of brewer's yeast that is similar to aroma compounds produced by ripening fruit. "The scent attracts fruit flies, which repay the yeast by dispersing their cells in the environment."[23]

These kinds of interconnections are why, Steve James explains, "we think of loss of yeast by association." *Saccharomyces cerevisiae* aren't airborne; they need a carrier, such as a wasp or bird, to get from one place to another.[24] "If a plant or animal goes extinct," Steve continues, "there is a good chance the corresponding microbes—including yeast—do, too." This diversity in yeast, or loss thereof, is a reflection of our actions and our history. Places like the NCYC give us an opportunity to save microbes (as well as plants and animals)—and all the smells and tastes they make possible—before they disappear.

"I think of the collection in two parts," Chris says. "Those yeast strains that are in use right now and those we might need in the future." Psychic predictions notwithstanding, the reason we can't know for sure what we'll need boils down to two factors. The first is that we might discover, somewhere down the line, that the thing we saved is vitally important—that a yeast cell has in its genes resistance to a disease or pest, or has a quality we don't yet know we need. The second is that only a fraction of the world's yeasts have been identified. It's not only hard to know what we have but, harder still, to know what we've *lost*.

Yeasts are immortal. They have been revived decades after they've been cultured. That's what enabled the NCYC to replenish the entire stock of Jennings Brewery in the Lake District in Cumbria after it was flooded back in 2009. "We had all the cultures they had lost in a safe deposit box. It took us three days to grow the yeast and get it back to them," Steve said. *Three days.* "And there's been no drop-off in viability from any samples in the last 45 years."

What's even more astonishing is that the entire NCYC collection fits into a vat of liquid nitrogen roughly the size of two kegs. Aside from the "Choose Chlamydia Screening" flier on the bathroom wall, this was the most surprising part of my visit. Intellectually, I understood microorganisms were micro, but this didn't really sink in until Chris donned thick blue gloves and lifted up the lid on the metal storage tank. I stood up on tiptoe and peered down into the icy chamber kept at about minus 196 degrees Celsius (minus 320 degrees Fahrenheit). Chris reached in and pulled out a long strand of freeze-dried samples stored in a series of small steel boxes. Inside each box were sets of red plastic straws roughly the length of my thumb. That was it. The entire collection—there, in a metal drum that I (at 5 feet and 95 pounds) could fit inside.

Since the dawn of agriculture, farmers have raised crops and animals for specific reasons, selecting for certain traits within the varieties they cultivate. Yeast is no different. Although baker's and brewer's yeast come from the same species, the strains are now quite different. Baker's yeast is a strain selected for its ability to ferment and make bread rise; brewer's yeast—of which there are multiple strains—have been selected by brewers for a range of characteristics, including the yield of yeast buds, alcohol conversion rates, aroma and taste.

These yeasts are domesticated—as different from wild yeast as modern corn is from teosinte. While yeasts can be domesticated to exhibit certain traits, they continue to be influenced by the environments in which they exist (like every other organism); they are constantly multiplying and changing. That is why the NCYC stores yeasts under stable conditions: Brewers (bakers or researchers) can always return to the master stocks of yeast held in the collection, knowing they'll be the same as when originally deposited.

Most yeasts, however, are still wild, carried on the surfaces of grains and fruits, as well as in the air, water and soil. Most brewers (bakers and winemakers) wash off or kill wild yeasts with heat and add domesticated yeast instead. This reduces variability and ensures their beer (bread or wine) will exhibit the characteristics required. But we need wild yeasts for the same reason we need wild relatives of corn, cacao and other foods. When an organism picks up or loses genes, it gains and loses various capabilities. By having access to raw genetic material, researchers can isolate the traits they want to keep.

Rather than shunning diversity, some brewers embrace it, actively gathering and using wild yeasts in their brews as an element of terroir (the taste of place). Belgian brewers, for instance, have made Lambic ales with wild yeast for over 500 years—a celebration of the rich biodiversity of the Senne river valley, a 24-by-120-kilometer (15-by-75-mile) region south and west of Brussels. They leave wort overnight in shallow pans called coolships, exposing the liquid to yeast and bacteria of the area.[25] Microbes in the air, along with the ones living in the wooden casks in which the ale is aged, start spontaneous fermentation.[26] Because the wort is cooled by the night air, Lambics were traditionally only made between September and April. In recent years, however, the production season has been shortened from October to March, a direct result of climate change.[27]

This isn't something we tend to think about when we're pouring a pint—that beer inputs have a deeper origin, that they have variability and seasonality. Beers' grains and flowers change in every season, with every harvest. Like any other crop, they're impacted by rain, sun, pests, diseases and myriad decisions made by the hands that grow them. Add to this the serendipity of fermentation and the ingenuity of brewmasters who turn a liquid, two solids and a microorganism into one of the world's most beloved beverages.

There are a few reasons we tend to think less about provenance in beer as opposed to wine. Beer includes multiple key elements beyond grapes and the yeast on grapes and in casks. And it gets its terroir from every one of its ingredients: water, grain, hops and yeast, plus whatever else might be added to the brew. Yet these ingredients— water, in particular—have often been tweaked to *lessen* the effects of terroir, rather than enhance it. For example, when Sierra Nevada Brewing Company expanded their operations from Chico, California, to Mills River, North Carolina, they added salt to the water to ensure continuity of taste. Oskar Blues Brewery, meanwhile, removed chlorine from the water at their nearby Brevard, North Carolina, facility, which is sourced from the municipal supply, as opposed to the run-off mountain water they collect from ponds at their flagship brewery in Longmont, Colorado.

This diversity—in water, in plants and in tiny organisms I couldn't even see without a microscope—led me to trace a yeast stored at the NCYC to what was once one of the largest breweries in the world: Truman's. The brewery was established in 1666 in East London on Brick Lane, a historic district at one time known for its breweries, then for its Indian restaurants, and now for its growing influx of nightclubs scattered between takeout curry joints.

At its height in the 1850s, Truman's brewed nine different porters and stouts. It exported beer to the West Indies, North America and Australia, and housed a stable of 200 horses for local deliveries.[28] Truman's was the first British brewery known to have employed a chemist—two decades before Louis Pasteur identified the role of yeast in fermentation.[29]

However, despite its popularity, beer drinkers' tastes evolved and, by the 1970s, Truman's beers started to fall out of favor. "Sadly, the jokes were no longer about the ability of Truman's beers to put you on the floor. Instead drinkers were asking what the difference was between Ben Truman and a dead frog, and giving the answer: 'There are more hops in a dead frog.'"[30]

In 1977, the company tried to catch up to a market that had grown to prefer imported lagers, but their new formulations (featuring different yeasts, malts and hops) never really took hold. Truman's then tried to return to its core offerings, but it was too late.[31] In 1989, after more than 300 years in existence, the British institution and cornerstone of British brewing history ceased operations.[32] The beers—and the diversity of flavors within them—disappeared.

The brewery remained closed until two local investors bought the Truman's brand in 2010 and relaunched it three years later.[33] Truman's is now located in Hackney Wick, a former industrial neighborhood in East London with the best street art I have ever seen.

I smelled the brewery before I saw it. The heady scent of ale cut through the cold London air and drew me into a cavernous facility that reminded me of an airport hangar. I arrived late to my appointment with Ben Ott, then head brewer at Truman's, the man tasked with bringing its beers back to life. "To resurrect the brews, the story," and, as he described it, "to see the Truman's eagle [its mascot] fly again."

Ben is a teddy bear of a man with shining greenish-hazel eyes, a big smile and tousled brown hair. He is also a complete science geek, as reflected in his extensive education in food science and technology—and the tattoo on his forearm of the chemical formula for linalool, one of six compounds that directly contributes to hop aroma.

"Before a drop is brewed," writes our trusted guide Randy Mosher, "the brewer must decide what's going in the beer. How strong will it be? What will be the color? Bitterness? Primary flavors? Balance? Sneaky, subtle background elements?"[34] This is what Ben does—and does well. A beneficent German man, Ben didn't chastise me for turning up over an hour late (I got lost); instead, he offered me a warm beverage to take off the chill after my long meander. The mug he handed me was full of sweet wort, a blend of water and malt.

Beer is 90 percent water, a liquid whose molecular composition and mineral content greatly influence the brewing process and flavor.[35] When water is destined for beer, we call it "liquor" to reinforce its specialness. Malt is dried grain that's been allowed to sprout, then dried again, then roasted. To crack the husk and increase the surface area of the grain (which makes its starches more accessible), it's run through a mill, then mixed and steeped slowly ("mashed") with hot water (liquor). In order to convert the starch in the malt to sugars, the liquor and malt are heated at set intervals. The mushy solids, known as mash, are then strained out of the liquid and the remaining broth is boiled to kill any unwanted microbes.

The wort Ben served me was golden brown. It was nutty, grainy and sweet, not unlike Grape-Nuts cereal. Not bad—but definitely not beer. Once I warmed up, Ben began to show me around. "Be careful," he cautioned. The floors were slick with beer and cramped with equipment.

We started with the grains, the foundation of my mug of wort and the beer that would follow. The base for Truman's beers is an English variety of barley called Maris Otter—a specialty malt developed in the 1960s by George Douglas Hutton Bell, a plant breeder and former director of the Plant Breeding Institute at the University of Cambridge.[36] Founded in 1912 to reignite the United Kingdom's sagging rural economy, the Institute bred over 130 crop varieties in a span of about 75 years, enabling Britain to become self-sufficient in its need for wheat and barley.[37]

Maris Otter is a winter barley, intended to be sown in the fall and take root before colder temperatures set in. It was introduced to brewers in 1966, specifically for English cask-conditioned ales (beer brewed from traditional ingredients and matured in a cask). Maris Otter grains grow quickly and malt easily, meaning they absorb a lot of water and, soon, start sprouting.[38] The malt also has "high diastatic power"—greater potential to convert the starches in the grain into fermentable sugars.[39]

Yet, despite these valuable brewing properties, Maris Otter was taken off the U.K. government's list of recommended malting barleys in favor of newer varieties with higher yields.[40] "By the 1990s no one was growing the barley at all," writes William Bostwick in *The Brewer's Tale: A History of the World According to Beer*. "What grain stores were left in the few old-timers' barns was all that remained, the last aromatic breath of a golden age."[41] Fortunately, the story doesn't end there. A consortium of farmers and malt producers (known as maltsters) bought the rights to market the heirloom seed and have saved it from disappearing.[42]

Ben scooped a handful of the plump grain out of a white plastic bag and dropped it into my palm. I was hungry—and it was delicious. The malt tasted like a nutty biscuit. Ben explained the nuttiness came through in the ale, and that Maris Otter also gave the beer a richer color and more robust body. Because Maris Otter is an heirloom, its yields are low (1.5 to 2 tons per acre compared to the 4 tons modern varieties produce) but its output is treasured. It is why, as one brewing website explains, "Craft Brewers around the World [sic] are willing to pay a premium for the complex, authentic pub flavor that Maris Otter provides."[43]

The malt is especially well suited to English IPAs (India Pale Ales), which is partly why it's been one of the most enduring varieties in the history of modern cereal cultivation. But the grain is now facing a threat much larger than the market shift to high-yielding varieties: climate change. All winter barley, unlike spring barley, needs a period of cold weather to trigger germination (the sprouting of a seedling from a seed). Warmer winters compromise flowering time, which is "tightly regulated by the plants' ability to measure the length of a day."[44] Like Lambic ales, whose production season has been shortened from October to March, IPAs are also suffering the effects of climate change.

As we walked from the mash tun (pronounced "mash ton")—the insulated vessel where the grains and water are mashed—toward the hops, Ben told me he was from Cologne "with some roots in Liverpool." (That was evidenced by a hybrid accent that occasionally confounded me: My initial notes on Maris Otter read "Mari Sotta.") His words were punctuated by the faint sounds of trucks backing up to the brewery and the loud banging by Ben's colleagues, busy hammering plastic plugs into the casks to seal them shut.

Hops—more specifically, common hops—are a natural preservative. Used by breweries for bittering and aromas ranging from floral to minty, they are often referred to as the "spice" of beer. Bittering hops are added to wort to balance out the sweetness of malt sugars, after which time the wort is known as hopped or bitter wort. This bitterness stimulates our appetite, priming us to pair beer with food.

Truman's hops, Ben explained, are just as specialized as his grains and yeast. "The unfertilized feminine flower is what we want for brewing," he said as he handed me a few dark green pellets that looked like rabbit food. He told me to roll them between my hands and inhale. The aroma was spectacular, a combination of sweet citrus and green tea. Some hops serve dual purposes and are used for both aroma and bittering, but Ben prefers specific hops for each part of the process.

Bittering hops are boiled in the wort for about an hour. After that, flavoring hops are added and then, at the end of the boil, a large handful of flowers are thrown in for aroma. Although we call them flowers, hops are technically strobiles—the cone-shaped fruits of the plant. The glands on the cones contain the bitter acids and essential oils that give brewers the qualities they seek.

Hops likely originated in China, the same place as *Saccharomyces* yeast. Researchers made this estimation based on the fact that China is the only country in the world where all species of hops are found.[45] It is not only the hop center of origin but also its center of diversity, the place where the most genetically diverse range of hops can be found.

The first documented use of hops, however, was in the eighth century when Benedictine monks used them for brewing in a Bavarian abbey outside of Munich.[46] Before hops, beer was flavored and preserved with gruit, a combination of heather, mugwort and other locally grown herbs and spices. The change was a bit of a tough sell. Influential Christian mystic and naturalist Hildegard of Bingen is believed to have written, hops "were not very useful. [They] make the soul of man sad, and weigh down his inner organs," while British physician and beer aficionado Andrew Boorde claimed hops made men fat and bloated.[47]

In that same spirit of skepticism, European breweries have historically shied away from American hops. In fact, Truman's was the first brewery in the U.K. to use them. It was a decision borne out of necessity when, in 1816, Britain suffered through a bad harvest.[48] But now it has become something else: "At first, people didn't like the piney taste of American hops," Ben told me. "Now they love it."

The diversity of hops reflects a diversity of tastes and traditions that are part of an extraordinary evolution in beer—particularly in the United States where American-style lager once defined beer in much the same way Folgers defined coffee. The image of American beer, the Brewers Association explains, "was simply that of a mass-produced commodity with little or no character, tradition or culture."[49]

When I was growing up, long before I drank from my first plastic cup of Bud Light, I remember beer marketers imploring beer drinkers to "Lose the carbs, not the taste." Stores and bars were saturated with light, low-calorie lager and little else.[50] The light beer explosion helped grow Big Beer and, by the end of the 1970s, industry experts predicted there would soon be only five brewing companies left.[51] (This drop was also rooted in earlier history, a product of Prohibition when more than 800 breweries closed their doors.[52]) Randy, our beer guide, writes: "The trend toward light, pale beer reached its low point with the introduction of Miller Clear in 1993. This water-clear beer, stripped of all color and much of its flavor by a carbon filtration process, was, thankfully, a step too far."[53]

Commercial beer, like commercial coffee or chocolate, is about the consistency of an experience. "We forced the diversity out of our food system," beer expert Garrett Oliver explains, "and we did it on purpose. It was done for commerce, so that one bland, long-lasting, well-preserved version of almost every food could be sold to us using mass advertising. And, with that, the memory of real food faded." This is why a Corona—or the Taco Bell 7-layer burrito that might accompany it—tastes the same in Dallas as it does in Seoul. It's not necessarily because the ingredients are the same, but because they have been modified to taste the same, year after year. And, in the case of Corona, beer after beer.

The two main yeast varieties used in beer also contribute to consistency in flavor; yeast is what separates the ales from the lagers. Lager yeasts ferment at cooler temperatures and drop to the bottom of the fermenter when they're done. Appropriately known as bottom-fermenting yeast, lager yeasts produce clean and crisp beers, like Corona, Heineken, Bud and Pabst Blue Ribbon. They are considered more commercial because they're uniform, controllable and don't produce the depth of flavor we find in ales. "If you want to attract a lot of people," Ben Ott explains, "then you make the beer as bland as possible." It works: Lager is the most popular beer in the world.

Lager yeast is a hybrid of *Saccharomyces cerevisiae*—the domesticated yeast used for millennia to leaven bread and ferment ale—and *Saccharomyces bayanus*, a species of yeast that gives beer the capacity to ferment at cold temperatures, creating a less cloudy brew. Lager yeasts are more predictable and offer less diversity than ale yeasts, not only in taste but genetics.

It makes perfect sense for companies to create beers that appeal to large audiences—and for us, the drinking public, to want something familiar. It's reassuring to be able to go anywhere in the world and have consistency in our favorite beverage. It's easy and safe.

But, in some ways, it's almost like going nowhere.

"What's better than beer?" one retailer asked. "Cheap beer!" But value is different from price. We're getting what we pay for. Is cheap beer—inexpensive sameness built on cheap labor and cheaper inputs—really what we want? In today's rich, complex world of beer, can we reach for something more? That's what a small group of brewers who had less interest in light lager sameness decided to explore, sparking a taste revolution that has transformed beer culture.

II. THE SOUL

Back in 1980, a burgeoning movement of craft brewers started evolving away from tasteless lagers to beers that more closely resembled European varieties. "The lack of a living beer tradition worth preserving," Randy Mosher details in *Tasting Beer*, "left [the United States] free to build a new beer culture from scratch."[54] The primary reason we lost diversity in beer—changing taste preferences—has now become the route to reclaiming it.

This effort included then up-and-coming American brewer Sierra Nevada, which released a hoppy pale ale made with domestic Cascade hops. *Smithsonian* reporter Natasha Geiling writes, "Cascade, and Sierra Nevada's Pale Ale, essentially started a brewing revolution, proving that hops with bitter, fruity qualities could produce a beer that sold well."[55]

Those hops offered a taste of place distinct from European (Old World) hops; they're genetically unique varieties with very different flavors and stories. Old World hops are reserved and earthy; they have been in Europe for over 1.5 million years and include some of the oldest, most traditional varieties of hops, known as noble hops.[56] Noble hops are highly aromatic and bring a subtle bitterness to beers; they are as prized and as geographically specific as a sparkling wine from the Champagne province in northeast France (the only place that can call its effervescent wine "Champagne") or sweet Vidalia onions, which can only legally come from Vidalia, Georgia. Only four hop varieties are *truly* noble—and only when they're cultivated in the areas in Germany and the Czech Republic where they are traditionally grown.[57]

American hop varieties, on the other hand, reflect American spirit: There's nothing subtle about them. They are intense and varied, known for being bright, citrusy and resinous. A number of these varieties can be used for both aroma and bittering, but they are best for bittering, as they tend to have higher concentrations of the alpha acids that are largely responsible for beer's bitterness.[58] While they are well suited for all pale ales, they have become a defining characteristic of American craft beer, especially American-style IPAs.

IPAs were developed in the 18th century when the British colonized India. There are multiple versions of how the pale ale became hoppier and more alcoholic, but suffice to say the Brits wanted their beer, so they tweaked it to better withstand the grueling passage from England to India.[59]

Food writer Maggie Dutton does the most interesting job of describing the English–American hops divide: "On the tongue, English-style IPA feels much the same as a strong black tea that has been brewed too long: Your taste buds will feel like suede rubbed the wrong way," she writes. "With an American-style IPA, you're likely to think tiny kittens have just skidded across your tongue, claws blazing, leaving your mouth scoured of all but the hint of hop."[60]

"The American hops we feature are very green and lychee," brewmaster Ben Ott explained as we walked through the brewery. The hops used at Truman's are grown in the Cascadia region of the Pacific Northwest—a 30,000-acre expanse across Washington, Oregon and Idaho (that also extends into British Columbia, Canada). Not only is the personality of these hops decidedly American, so is its production. "The hop industry—though outwardly sexier than corn or soybeans—is still a product of modern industrial agriculture, where centralization and tradition reign supreme," Natasha Geiling explains. "The United States produces nearly one-third of all the hops in the world—of that, 79 percent is grown in Washington state. Nearly half of all hop varieties grown in Washington state fall into four hop varieties: Zeus, Cascade, Columbus/Tomahawk and Summit."[61]

Growing a limited range of crops, as we know, increases risk, including vulnerability to disease. For hops, most of the danger lies in two fungi: downy mildew and powdery mildew.[62] And, like the witches' broom that chokes cacao, there's no known cure.[63] Farmers have been instructed to manage the pathogens by cultivating disease-resistant varieties, pruning plants, applying fungicides, and killing any wild varieties of hops that could be possible carriers of disease.[64]

But those wild hops might also include varieties that are *resistant* to diseases or other menaces—or expand the diversity of flavors the market craves. It's why Todd Bates and Steve Johnson, organic farmers from New Mexico who established one of the first hopyards in the area back in 2002, are trying to change the "kill wild hops" mandate.

Todd has been curious about the medicinal properties of plants since he was a kid. A child of the '60s, he started collecting wild hops in northern New Mexico that were so distinct from the ones grown in other parts of the United States they were given their own taxonomic designation (a distinct variety of common hops called *neomexicanus*).[65] But when he and Steve decided to dedicate a portion of their land to growing them—and asked neighboring farms to do the same—people thought they were crazy. "The response people gave me was 'Why? That shit grows all over my fence. Why would I want to grow it?'"

Farmers weren't the only ones questioning Todd's sanity. "I went into a meeting with Ralph Olson, the CEO of [Washington-based] craft hop supplier Hopunion, and he was really nice, but I could tell I was being treated as the goofy guy who was a little touched. And then I got it. I was in a place surrounded by signs telling people to eradicate all wild hops."

Researchers cautioned against any experimentation with wild hops, Todd said, because of "500 years of people saying no one would drink beer made from them." Venturing out into the great (wild) unknown had real financial consequences for farmers and brewers. Growers had no desire to cultivate wild varieties that most considered weeds and had none of the sensory properties brewers were looking for. Todd was at a standstill, but he knew he had something special. His hops thrived in the worst of drought. "And they had crazy, psychotic vigor. But the term 'wild hop' was infectious. No one wanted to touch it. I just meant hops from the mountains—pure American hops."

These varieties thrive in challenging places and offer flavors that aren't necessarily unpalatable, just unfamiliar. So Todd teamed up with hop farmer Eric Desmarais to identify what brewers would want. Eric runs a family hop farm in Moxee Valley, Washington[66]—one of three distinct growing areas in the Yakima Valley that contains about 75 percent of the total U.S. hop acreage.[67] He had already developed the El Dorado (a hop known for its tropical fruit flavors) and was eager to explore further.

Todd gave Eric 80 varieties, which Eric then narrowed down to two he thought would make good beer. One of them, Medusa, made its national debut in Sierra Nevada's Harvest Wild Hop IPA series. (Sierra Nevada is now the third-largest craft brewery and seventh-largest brewery in the country.[68]) The company explained, "These bizarre, multi-headed, native U.S. cones have a flavor like nothing we've tasted, and for the first time, we're showcasing their unusual melon, apricot and citrus aromas and flavors in our beer."[69]

Medusa and other local hops have the potential to not only transform craft beer but also reshape the entire brewing industry. Native to America, their hardiness might provide an advantage against global warming and allow growers to expand into places that haven't had much success cultivating the plant—ranging from San Diego to the mountains of New Mexico.

Diversity in hops reflects a diversity in tastes and traditions that craft brewers in the United States are bringing to the fore. Craft beer is small, independent and traditional. According to the nonprofit Brewers Association, in order to be identified as "craft," two-thirds of a brewing operation has to be owned by craft brewers, with an annual production of 6 million barrels or less of beer (not flavored malt beverages).[70]

While fine chocolate is gaining traction and specialty coffee is expanding, craft beer has been on a steady growth trajectory since 2003. The sector nearly doubled between 2007 and 2012 (from $5.7 billion to

$12 billion)[71] and, in 2014, succeeded in edging out the self-proclaimed King of Beers, Budweiser.[72] Craft beer is forecasted to grow into an $18 billion industry by 2017[73]—a far cry from the 1980s, the era in which I was introduced to beer.

My beer preferences, along with a growing number of Americans', have evolved away from Bud Light and toward New Belgium. Sales of craft beer have doubled in the last five years, while sales of light beer are projected to hit a 10-year low.[74] But make no mistake: Those who love Bud are legion. "Light, dark, strong, weak, fizzy, flat, canned, bottled, or draft, beer has fluidly adapted to serve every role it has been asked to play," Randy reminds us. "Beer is the universal beverage."[75]

Sustaining consistent flavor in beer (and other foods and drinks) requires a constant tweaking of ingredients that, by their very nature, are changing. This happens in conjunction with palates that are also changing, so subtle differences aren't easy to detect. Take bitterness, for example, measured in International Bitterness Units (IBUs). "Sierra Nevada Pale Ale has an IBU of 38," Ben explains. "It used to be one of the hoppiest beers in the world. Now people say it tastes less bitter. But it hasn't changed, we have. We've gotten used to the bitterness. Our taste buds have changed over generations."

We're now experiencing a "hop rush"[76] and the introduction of beers that are so bitter they exceed 100 IBUs (the maximum amount of bitterness humans can detect). Food and beverage companies count on this evolution. We aren't born loving beer or coffee; we grow to enjoy them. Although our taste maps feel stable ("I don't like beer"), they're always changing, which is why the people behind the foods and drinks we love are in a constant state of recalibration. Companies introduce (and foster through fierce marketing) new flavors—while also doing their best to sustain the tastes to which we are accustomed.

But hops, of course, do more than bitter. They also add complex aromas, which is another reason craft brewers have embraced them. Mark Dredge, author of *Craft Beer World: A Guide to Over 350 of the Finest Beers Known to Man*, designed the Beer Flavor Wheel (on page 349) to highlight these emerging smells and tastes.

The wheel, like all wheels, is a cultural construct reflective of current trends in beer. The original beer flavor wheel, developed in 1979 by Danish chemist Morten Meilgaard and his colleagues, was primarily created to identify flavors in pilsners and lagers.[77]

Mark felt the original wheel was "too scientific and not approachable for the regular drinker, there to critique beer rather than appreciate it." So he started thinking about how he could make the information more accessible. "I wanted something that immediately helped the drinker to recognize and reference what they smelled and tasted in a positive way. I also wanted there to be some complexity to it so that people could understand that smelling strawberry in the beer is a yeast-derived ester, whereas something like grapefruit comes from hops. Plus, I thought it was important to consider things like mouthfeel to help improve the overall language used for craft beer appreciation."

In craft beer, what was old is new again—an attempt, in both ingredients and brewing techniques, to return to the origins of what makes beer special. "We're going back to our roots," brewmaster Garrett Oliver told me. "It feels like a new invention, but I say to my fellow brewers, 'Get over yourselves.' People have been brewing beer for over 20,000 years. We forgot almost everything—and now we're remembering."

This same spirit of remembrance is what inspired Ben to bring back Truman's traditional brews, the foundation of which is its yeast. "Brewers make wort, not beer," Randy underscores. "Yeast makes beer."[78]

As Ben and I walked from the hops over to the fermenter, the air got perceptibly warmer. The transformation from wort to beer was imminent. Through the process of fermentation, yeast digests sugars from the starchy malt and creates ethanol and carbon dioxide. This starts, Randy explains, when "the yeast takes in the oxygen and begins to make more yeast by 'budding' off new cells. This takes several hours, and during this time there is very little actual fermenting going on. But at a point, all the available oxygen is used up and the yeast turns its attention to the sweet wort. First, because it's easier, the yeast eats the small amount of available glucose (a single sugar), and then it begins to metabolize the [malt sugar] maltose. This tiny ravenous beast can throw a thick, rocky head on the surface of the fermenting beer more than a foot high and generate enough heat that tanks need to be cooled to avoid runaway temperatures."[79]

After the yeast has done its work, the leftovers are collected and used again (repitched). Ben reuses his yeast about 60 to 90 times. "There's nothing special about my Maris Otter," he says. "And the water? Well, I have to live with London water. And the yeast, I could just order the powdered stuff and it will be drinkable and it will be fine. But 1126 underlines my beer perfectly. The yeast gives it its soul."

Yeast 1126 is what Ben retrieved from the National Collection of Yeast Cultures—it's what brought his beers back to life. The strain Ben retrieved is one of six that Truman's has stored in the NCYC collection. Steve James says, "A lot of the group that Truman's drew from in trying to bring back one of their original brews were deposited here in the late 1950s and '60s in what we call 'the open collection with confidentiality.' That means a number of people could access it, but, unless they had the strain number, they wouldn't know what brewery the yeast came from."

Those yeasts reveal different characteristics depending on the brewer and inputs he or she uses. "Craft beer is very hops-driven right now," Ben added. "They don't worry that much about yeast. But we wanted to bring the Truman's name back—and that meant we needed to get the yeast."

For Ben, Truman's future lies in its past. The strain 1126 is a link between a historic ale and the modern-day return of the Truman's eagle. He is the first brewer in 21 years to use this yeast. "We wanted one that would work well—that performs a good head. The younger yeast wouldn't do that. So this provides a competitive advantage," one that's accomplished by bringing back the yeast, the brewing tradition and the story. "The U.K.," he said, "is losing big pubs but growing in microbreweries. This is our drive for difference in flavor and taste."

And it was a drive for something else. What I understood, as I wound my way through the brewery with Ben, was that this wasn't simply about resurrecting a yeast or bringing back a beer. It was a connection to the bigger story of Truman's, East London, Brick Lane and the whole history of beer. Beers made with yeast 1126 were a way to honor the lineage of brewers who had come before him—and those who would come after. That is what those pints Ben poured represented; they held the story.

In each of my journeys, there has been a moment when everything I was struggling to understand came into focus. With wine, it was when I touched Peter Fanucchi's grapevines; in coffee, it was standing in the Kafa forest with Mesfin Tekle. I had been working all morning to appreciate beer; it was an effort. And then—as I stood in sneakers sticky with beer, eating barley, drinking wort and smelling bready yeast—my appreciation for beer suddenly became effortless.

Ben poured me (the woman who nurses tiny beer samplers) multiple pints. I was sitting on a wooden bench, close to a shaft of light that illuminated the beers from behind. They were beautiful—rich hues of red, yellow and brown I had never been able to see in dim, crowded pubs. Each pint was topped with creamy peaks of foam I wanted to scoop up with my finger.

I hadn't taken a sip, but I already felt nourished.

"Beer is a total package," says Charlie Bamforth, the professor of brewing science at the University of California, Davis, who's popularly known as the "Pope of Foam." "It has an aesthetic and a taste, but color and foam are the most striking things people first notice."[80]

We first taste with our eyes. The appearance of beer, or any beverage or food, has a halo effect that impacts our willingness to sample the product. Is it pretty enough? Fresh enough? Delicious enough? The eyes decide—and inform our ideas about what the food or drink should taste like.[81]

In beer, this visual assessment boils down to what I like to call the three Cs: carbonation, color and clarity. Carbonation manifests as beer foam, lace and the tiny bubbles that work their way through the drink. They offer great clues about flavor and mouthfeel.[82] Color (what Randy describes as a "whole rainbow"[83]) might indicate the strength of a beer but is more likely a reflection of beer type (think of a dark brown oatmeal stout versus an amber wheat beer). Clarity can also be a reflection of style. Some beers, including Ben's bottled-conditioned beers, are meant to be hazy, others crisp. They are all signs of what's to come.

When I fumbled with the small, clear vials in the Le Nez du Vin kit as I tried to make sense of wine flavors, I floundered not only because I was tired and sad, but because every vial looked the same. They were all filled to the same level, sealed with black caps and numbered for identification. As one study put it, "The nose smells what the eye sees."[84] What we see is what we understand: I needed to *see* what I was smelling so I could identify it. The wine students in Bordeaux faced similar challenges when they used erroneous red wine descriptors to describe tinted white wine. This was also at the heart of the horsemeat scandal that rocked Europe back in 2013. People happily ate burgers and frozen lasagna they thought were made of cow, but they were actually horse.[85]

Visuals matter.

The primary organ of sight is, of course, the eye. But, as with all other senses, in order for what we see to register, we also need our brain—and the nerves that transmit messages from the eyes to the brain. This is why "seventy percent of the body's sense receptors cluster in the eyes."[86] Our eyes are fluid-filled balls of water and collagen (the most abundant protein in our bodies[87]), plus muscle. They sit in the hollows of our skull, protected by our eyelids, lashes and the bony protrusions of our eyebrows.

The eyeball is covered with a clear, curved cornea, which rests over the iris—a colored ring on the white of our eye that gives it its color and contains muscles that cause the pupil (the dark circle in the center of our eye) to contract or enlarge.

The pupil is the place where light comes into the eye. It changes size depending on the availability of light (or the intensity of, say, a marijuana high). Just behind the pupil is the lens, a clear, disk-shaped structure that bends rays of light and shines images onto the retina in the back of our eye. The retina is a delicate layer of tissue roughly the size of a postage stamp that's filled with about 130 million light-sensitive cells known as photoreceptors. It is the place where visual processing begins, where light rays turn into images. If you imagine the eye as a camera, our lens is the aperture and our retina is the film.

"Color," author Diane Ackerman states, "doesn't occur in the world, but in the mind."[88] It is both a physiological and psychological response to that narrow band of light frequencies we call the visible light spectrum. When pondering the hues in beer or trying to make sense of a stoplight, this interplay between the eye and brain is critical. It's what tells us to stop or go.

The spectrum of color in beer was originally identified through a set of stained glass that measured the wavelengths of light in beer and was created by British brewer Joseph Lovibond. Today, American brewers use the American Society of Brewing Chemists color standard, called the Standard Reference Method, or SRM.[89] (Europeans use a different scale.)

Our perceptions of those and other colors are about what we experience in the moment and what we *expect* to experience in the moment. That's why Miller Clear and Crystal Pepsi flopped: Our brain had an expectation that beer should be amber and soda should be brown. It's also why large-scale brewers have strict color specifications for their finished beers and sometimes blend batches or use roasted barley to reach the shade they're looking for.[90] They want to meet our visual expectations because they know that the initial assessment will inform everything that follows. Craft brewers have typically placed less of an emphasis on color analysis, but they also want those perceptions to line up.

In one study exploring the misalignment of taste and color, scientists served 59 college students at the University of Florida four cups filled with pulp-free orange juice. They adulterated some cups with flavorless yellow food dye and Equal sweetener to create varying intensities of color and sweetness. When participants were given two cups of orange juice with one cup darkened yellow, they perceived taste differences that didn't exist. But when they were given two cups of orange juice that were the same color but one cup had been sweetened with Equal, the students missed taste differences that actually *did* exist. The researchers summarized: "The visual cue dominated the taste cue." The eyes overrode what the tongue tasted.[91]

Many studies show similar results. My favorite (likely because of my Le Nez fail) was a small study of 11 women between the ages of 25 and 42 who were given transparent cups filled with distilled water tinted with flavorless dyes. The women were asked to match the colored liquids to a list of flavor descriptors, linking the colors to what, in their minds, seemed like logical flavors. Most associated the orange liquid with orange flavor, the yellow liquid with lemon, the blue solution with spearmint and the green one with lime. The grey solution was considered a toss-up between black currant and licorice, and the red was estimated to be strawberry, raspberry or cherry. Although the liquid was just colored water, and the women were simply looking at the liquid, not tasting it, they were already primed. They believed certain colors had strong connections to specific flavors.[92]

The researchers then flavored the water with essences of strawberry, cherry, lime and orange, dyed them appropriate and inappropriate colors, and had the same group of women taste them. When the cherry water was colored red, everyone identified it as cherry. But when the water flavored with cherry essence was dyed orange, 40 *percent* of the women thought it tasted of oranges.[93] This mismatch continued with every cup of colored and flavored water with one exception: The strawberry-flavored liquid never confused the volunteers; they consistently identified it as strawberry, regardless of its color.[94]

These responses are primal. Evolutionarily speaking, they are another way our bodies try to keep us from ingesting poison. But these assessments aren't universal—at least not in beer. Researchers have discovered the most influential visual component in beer is foam.[95] Studies on color, beer foam (or head) retention and lacing (foam remnants stuck to the side of the glass) show that Americans and Italians both want Goldilocks foam: not too much, not too little, just right (which, it turns out, is roughly 1 inch).[96] However, there are differences in what they perceived foam to represent. For Italian women, "huge foam and lacing were synonymous with freshness, fizziness and thirst quenching."[97] But in a study done with Americans, beers with bigger heads and more lacing were believed to be less thirst quenching than beers with moderate to low levels of foam.[98] In addition, "the beer with the highest head and lacing ranked as having a dirtier glass than the beers with moderate levels."[99] This notion is incorrect: Dirty glasses *destabilize* foam.

Foam is a product of hop residue, plus proteins found in grain and yeast. It consists of bubbles containing carbon dioxide gas, which yeast gives off during fermentation.[100] The common belief in my high school beer-drinking days was that more foam equaled less beer equaled a diminished chance of enjoyment. In actuality, "the quality of the foam in any given glass of beer is the result of many ... factors: how the beer was brewed, the ingredients used, the level of carbonation, the serving glass and even how the beer was poured."[101] A thick, creamy head and prominent lacing are signs of a good beer. Yet, right or wrong, perceptions that beers with less foam are less thirst quenching are some people's realities—at least until they taste the beer.

Transmission and interpretation of visual messages are nonstop. They start from the moment we open our eyes and end only when we close them to sleep. This processing of shapes, movements and colors guides our engagement in and with the world; it's the primary response system that helps us decide whether to get closer to something (a beer, a potential beau) or move farther away.

The eye is the same organ that makes tears, simultaneously protecting itself and revealing our joys and sorrows. Eyes are the windows to the world and our soul. But while we take in that world, we have to remember these perceptions can change, especially when we slow down and pay attention to what's normally an unconscious transmission from eye to brain.

As author Anaïs Nin once wrote, "We don't see things as they are, we see them as we are." Beauty is in the eye of the beholder. We have the ability to supersede our first impressions and create new ones—learning to respect beer and all other sights we take in, and staying open to, some day, loving them.

The last thing Ben said as he passed me the pints was that every beer at Truman's uses yeast 1126. I stared at the tall, slightly cloudy glasses. That yeast ... I had touched that yeast in little red straws and traced the strain all the way to this moment. I had chewed the malt and smelled the hops and was now ready to taste what they had become. Truman's Runner, the brewery's flagship ale and the first Ben reintroduced, is reddish copper in color and has a light citrus nose. The flavor is wonderfully malty, with chocolate aromas in the front and hints of citrus and bitterness at the back, the result of their signature blend of traditional European hops. Truman's Swift—a significantly lighter golden brew that uses American Cascade and Czech hops—was, true to form, bright and crisp. The final beer, a limited edition Tom Ditto American Pale Ale, also used a blend of American and European hops. It was my favorite: fragrant and doughy, simultaneously bitter and sweet.

Not surprisingly, I couldn't finish the pints. But I liked them. For the first time in my life, I *liked* beer.

After I drank my lunch, I poured myself a cup of coffee to ward off the chill. When I requested sugar, Ben presented me with a spoon of sweetened barley malt and a wink. "We're in a brewery." As I nursed my coffee, I watched Ben's colleagues, Alessandro Reali and James Rabagliati, prepare the metal casks destined for pubs across England. They were kitted in tall rubber boots, multiple layers of shirts and woolen hats—hoisting and filling the metal casks with beer that made the floors both slick and sticky. They hammered in the plastic plugs on cask after cask. The sound was deafening: *Bang! Bang! Bang!* Yet, despite the wet floors, cold weather and insanely loud noise, there was precision and ease in how everyone worked.

In a keynote at the 2006 Craft Brewers Conference, Dogfish Head Craft Brewery founder Sam Calagione stressed, "Breweries have terroir as well. But instead of revolving around a patch of land, ours are centered on a group of people." To which beer author and amateur brewer Stan Hieronymus adds, the way terroir "expresses itself in a particular beer remains within the province of the brewer. That hop could be blended, then balanced against malt and fermentation character, becoming part of a greater whole. Or it could be showcased as the only hop in an appropriately named 'single hop' beer, dominating the aroma and flavor."[102] Terroir in beer is as much about people as it is place.

Once my coffee turned cold, I decided to wrap things up. It was nearly 3 P.M. Ben, Alessandro and James had been at work since 6 A.M. and wouldn't stop until closer to 4, but after five hours of wandering around and fitting myself into odd corners to jot down observations, I was cold, tired and ready to head back to my hotel.

As I snaked my way through Hackney Wick toward the train, I started to wonder if maybe the experience of beer was just too masculine for me—if beer and beer culture were made for men, and men alone. This was before I learned that beer is as feminine as it gets—from the grain to the pub.

Since the 1950s, archeologists have been exploring the idea that the dawn of human civilization was actually based on the cultivation of grains for beer, not bread.[103] But even before we started sowing and reaping, we were hunting and gathering. Our female ancestors— the gatherers—collected grain long before they started growing it.[104] Researchers speculate the earliest brews were likely happy accidents, the result of rainwater entering vessels used to collect barley and other cereals. The soggy grains sprouted and produced sugars, the wild yeast on the grains and in the vessel converted the sugars to carbon dioxide and alcohol and—voilà!—a thick, soupy beer was born.

"The 'high' that people obtained from beer," University of Pennsylvania anthropologists Solomon Katz and Mary Voigt explain, was "the number one reason for the domestication of wild grains."[105] Before we ended our day with beer, we started our day with it. We drank all day, every day—old and young, rich and poor. We evolved through beer, and beer evolved through us.

The earliest known alcoholic beverage was a 9,000-year-old Chinese beer-wine made from rice, honey and fruit,[106] but the first evidence of grain-based (barley) beer dates back to the Sumerians of Mesopotamia, the first civilization in human history. Researchers verified this timing through ceramic vessels used for brewing and a clay tablet from the 19th century BC.[107] Inscribed on the tablet was "Hymn to Ninkasi," a poem dedicated to the Sumerian goddess of brewing and beer that included a recipe for brewing made by the goddess herself:

You are the one who holds with both hands the great sweet wort,
Brewing [it] with honey [and] wine

(You the sweet wort to the vessel)

Ninkasi, (...) (You the sweet wort to the vessel)[108]

The Sumerian word for beer—*kas*—translates as "what the mouth desires."[109] It was the cornerstone of the Sumerian diet and culture, divinely sanctioned and protected by Ninkasi (also known as Siris) and Siduri, early Mesopotamian goddesses associated with beer and brewing.[110] Paige Latham, amateur brewer and author of a web series on women and beer, explains, "For centuries, the *only* area of society in Ancient Mesopotamia that was sanctioned and protected by female goddesses was brewing."[111]

The people of Babylon, the civilization that followed the fall of Sumer, also considered beer a gift from the goddess, as did the Egyptians, whose deity of beer was associated with childbirth. Tenenit, whose name derived from *tenemu* (one of the Egyptian words for beer), was closely linked with Meskhenet, the goddess of childbirth and protector of the birthing house. Tenenit wore a cow's uterus as a headdress and passed down a tradition of brewing believed to have been taught by Osiris, the god of death, resurrection and fertility.[112] These gifts of beer and brewing were transmitted from goddesses directly to women.

Egyptian workers were often paid in allotments of beer, and that included those who built what would become one of the Seven Wonders of the Ancient World. "Beer, along with bread and onions, is credited with fueling vast construction projects like the Pyramids."[113] This was because beer was safer and heartier than water: "While safe, sanitary water was largely nonexistent, the bready porridges from which beer was made provided essential nutrients for maintaining health, and the alcohol provided antiseptic qualities and was safe from disease."[114] But those workers weren't just carb-loading; they were tipsy. The Pyramids were built by the grace of beer—and the women who brewed it.

You see, in those early civilizations, stretching from what's now Africa into Asia and Central and South America, beer was women's work. In Europe, "the ancient Finns believed three women created ale for a wedding feast by serendipitously mixing a bear's saliva with wild honey and blending it with beer. Cerevisia, the Romans' brew, was named for

Ceres, the harvest goddess."[115] And in England, women in beer came to be known as "alewives" or "brewsters"—women who brewed beer out of their homes as a way to earn money for their households.[116] Alewives' homes were the first brewpubs, places where people could buy ale and come together for a drink.[117]

Throughout the Middle Ages, women continued to be the main brewers and sellers of beer. This work gave women an income, social standing and, if married, a degree of independence from their husbands.[118] But once men discovered how lucrative the beverage could be (people, at the time, drank an estimated 1 gallon of beer per person per day), they took over most of the operations.[119] "Like in Scotland," Garrett Oliver explains, "where they created a law limiting the amount of barley women could buy." By the 1600s, most brewers in England were men (many of whom were monks[120]). This trend continued through the colonization of the New World; America's founding fathers George Washington and Thomas Jefferson were brewers.[121]

Women's role in beer transformed so much that, by the late 1800s, they were at the forefront of the movement to renounce alcohol. The push for abstinence, led by the Women's Christian Temperance Union, was, at its core, an effort to reduce poverty and domestic violence by limiting alcohol consumption. It was, ultimately, a means to empower women, who formed WCTU chapters all across the United States and joined together in anti-saloon crusades. The group, founded in 1874, became the largest women's organization both domestically and globally. By the early 1900s, women had sown the seeds for both the 18th Amendment—Prohibition—and the 19th Amendment, granting women the right to vote.[122] Continuing in the pursuit of sovereignty, women were also at the forefront of the *repeal* of Prohibition, a movement fueled by personal privacy and the belief that decisions about the consumption of alcohol were best left to the individual, not the government.[123] Yet, despite their great influence in transforming legislation, which included hidden and open enjoyment of alcohol, women's direct connection with the beer they had brought into the world has never been the same.

What women became weren't people who made beer or even really drank it; they became the ones who sold it. Among the earliest and most successful saleswomen was Mabel, a character created in 1951 for Carling Black Label beers. Mabel was "a genial blond bartendress who rarely spoke, but ended virtually every commercial with a friendly wink. ... One observer commented that Mabel could 'compel any man to leave home—to fetch a carton of Carling's, that is.'" Mabel's wink and smile helped catapult the Carling Brewing Company from the 28th to the 6th largest brewery in North America.[124]

Over the decades, this trend has gone from bad to worse, with ad after ad showing voluptuous women in bikinis either frolicking through foamy waves or serving men manly brews. When I was coming of age in beer, I couldn't find my place in this—in fizzy, light lagers or in commercials that told me to "grab a Heine."[125]

But there was also something else. It wasn't just the bad beer and sexist advertising; it was the perception of what I thought beer was: a less refined drink. I sat in my occasionally curry-scented house desperate to be accepted—to be a sophisticate. I would never be a debutante or go to the cotillions some of my friends complained about having to attend, but I wanted to.

In an attempt to belong, I lied to my friends about celebrating Christmas (we didn't; my family is Sikh and Hindu), permed my hair (which I then dyed a horrible shade of auburn) and donned green contact lenses. Of course that didn't work. Even I knew that. The perm grew out, the colored contacts were swapped for clear—but that yearning to belong endured.

On my first day of college, a senior named Diane (from a suburb of Boston, with an accent so thick I could hardly understand her) said, "I can't believe someone who looks like *you*, talks like *that*." What she meant was that someone who looked Indian would speak with a Southern accent. I lost the accent. I also stopped shaving my legs and armpits and started writing "womyn" with a *y* not an *e*. I was a budding feminist at a prestigious women's college; there was no space for m-e-n in the collegial sisterhood. I wanted to fit in.

Today, nearly a quarter century later, I speak with a nondescript accent that pegs me as American but nothing more. I remove all unwanted body hair with lasers and wax, dye my grey roots black, and spell "women" with an *e*. I am proud to call myself a feminist. And I am proud to live in a time when women are, again, redefining their place in beer.

In a 2012 Gallup poll, beer edged out wine as the favorite alcoholic beverage for women ages 18 to 34. Wine had long held the top spot (and still does with older adults, especially women) but has been displaced in the younger demographic due to the growing popularity of craft beer.[126]

"A lot of women don't think they like beer," said Kelsi Moffitt, coordinator of the Twin Cities chapter of Barley's Angels, a worldwide network dedicated to expanding women's appreciation for craft beer. "You have to be willing to try new things. There are some beers you're not going to like, and some you will. But instead of turning away, be open. Don't give up. Give yourself a chance to try."

There, in beer, another life lesson.

Barley's Angels is a community that's developed around a love of beer. "It was always a woman's job to make beer," Kelsi told me. "Now we're taking it back." About 40 percent of the women in the Minnesota chapter are home brewers. "As women, we're constantly battling these ideas that when we go to the brewing supply store we're picking up this stuff for our husbands or boyfriends. No. We're brewers—and we know what we're doing."

Kelsi concluded, "There's a primalness to making beer. It feels very natural and simple; it's like cooking. We've passed this down from generation to generation. We know how to do this, and it feels good."

This is what I wish I'd known as I worked my way through the streets of London back to my hotel. Beer was the preferred beverage of goddesses, women and men—in that order. Brewing was women's work, and brewpubs were invented by women. This drink, so revered by men, exists only because of women. In my youth, I hadn't wanted to be considered the kind of woman who drank beer. But that is exactly who I am, and that is exactly who I want to be.

III. TASTING BEER

Beer has a generous spirit, so if you make any kind of effort at all, it will reward you with a rich and memorable experience.

—RANDY MOSHER, *TASTING BEER: AN INSIDER'S GUIDE TO THE WORLD'S GREATEST DRINK*

As always, try to start with a clear mind and palate and remove any sensory distractions (candles, perfume or even lipstick, which interferes with the structure of beer's head). Make sure to have on hand water and plain water crackers so you can reset your palate between beers. You might also want a spit cup: The goal is still enlightenment, not inebriation.

Start where you are—or where your guests are. If you're accustomed to light lagers, begin with a selection of beers that are less hoppy and have lower alcohol content. If you prefer wine to beer, start with a fruit beer, such as a Belgian Lambic (the opposite of light lager). If your tastes tend more toward gin, look for hoppy IPAs with aromas of evergreen and juniper. Regardless of your selection, let the lighter, subtler beers lead the tasting. If you flood your senses with a dark oatmeal stout, you'll have a much harder time finding the nuances in, say, a pilsner.

Pull your beers of choice out of the fridge and let them warm to their optimal serving temperatures. I was raised to believe beer—like revenge—was best served icy cold, but I was wrong. The look, smell, feel and taste of beer is dramatically affected by temperature. "The antiquated notion of frozen glassware," explains beer purveyor Paige Latham, "is actually a detriment to tasting because the water freezes."

Unless it's the middle of summer or you're in the tropics, let your drinks sit at room temperature for about 20 to 40 minutes before you're ready to try them. The range for specialty beers should be between 38 and 55 degrees Fahrenheit. As a general rule, keep "stronger beers warmer than weak; dark beers warmer than light."[127]

Consider the vessel. Just as beers have optimal temperatures, they also have optimal glassware. A good all-around glass is a tulip-shaped beer snifter (or even a large, stemmed wine glass), which concentrates foam and captures volatile aromas as they waft from the brew. Strangely, most of us drink beer from a vessel experts consider suboptimal: the classic shaker pint glass. But if that's all you have, use it. The worst option is to keep all the wonderful smells and foam potential of beer locked up in a can or bottle. Get your brew into a glass—even a plastic cup—and you've already made it taste better.

Ensure your glass is clean and, ideally, used exclusively for beer. Detergents and other substances (like milk) have fats that can leave film on a glass. This can compromise foam and make beer go flatter faster. A paste of baking soda, salt and water will clear off any residue and make your glassware beer-ready.

Beer is a complex interaction of simple, accessible ingredients. Start your tasting with these elements. The Beer Flavor Wheel found on page 349 will help you describe what you experience, starting with water, the foundation of beer. Try tap water, distilled water, plus a few different types of bottled water. Recognizing these differences will help you better understand why brewers tweak water to get it just so. Try to also sample small amounts of hops and malt. You can likely procure them from a home brewer, local brewery or online. Roll the hops together between your hands, cup them and inhale. Smell and taste the malt.

And now pour the beer. The side pour to which we've grown accustomed is wrong: "Trickling down the side is for sissies and will result in a too-gassy beer with little aroma and a poor, quickly dissipating head."[128] Beer should be poured, just before tasting, right down the middle of the glass. The key here is patience. Go slowly, wait for the foam to settle and pour again. By allowing the foam to grow and shrink, the head will be creamy, dense and longer lasting. The patient pour also allows more of the carbonation to escape, which means you'll fill up on beer, not air. Foam traps airborne aromas. A nice fluffy, just-right head will enable you to experience a greater depth of flavor.

Pour until the glass is about one-third full (leaving room for the head and volatile aromas). While you're doing so, *listen.* What comes up for you as you hear the sound of a bottle twisting or can popping, as the carbonation escapes and the beer glugs into the glass? Notice how your body starts to prime itself to taste the beer.

"Once the beer is in the glass," Randy explains, "it's up to you to decide what your relationship to it will be: advocate or critic, stranger or best buddy. Like any worthy pursuit, the beer-tasting experience is only as good as you make it. Summon all your senses and experience and dive into it wholeheartedly."[129]

Look at the beer. Is there a chill on the glass? If so, wipe it off so you can clearly assess the color, clarity and carbonation. What shade is your beer? What is the size and density of its head? Is your beer cloudy or clear? Color and clarity are defining characteristics of certain beer styles. What signals is the beer already giving you about its taste? What expectations do you have about its flavor?

Make mental notes, write them down, or talk them through with your fellow tasters. Strengthen those neural connections.

Now raise the glass to your mouth and start to take in all the aromas. This journey is similar to ones we've taken with wine and coffee. Swirl the beer in the glass the same way you swirl wine. Draw in all the aromatic compounds in a series of quick sniffs, allowing the air to come in from both your nose and mouth. Notice what memories your olfactory system triggers.

There are hundreds of aroma and flavor compounds in beer that come from each of its ingredients, plus the alchemy that happens when the substances are combined, fermented and processed. Mineral ions in water can give beer a subtle saltiness, astringency or minerality. Malt imparts sweetness (and also bitterness) and smells ranging from caramel and bread to toasted nuts and coffee. These aromas can come from either the grain or the Maillard reaction in the roasted malt. Hops bring out bitterness that offsets the sweetness of the grain, but also offer their own sweetness and aromas.

The taste of hops is the one most closely associated with place. See in the beer wheel how hops from certain areas are connected to certain aromas. Now check the wheel for yeast. Yeasts, like grain, provide nutrients that make beer a formidable source of nourishment. They're responsible for alcohol and carbon dioxide and also tastes and smells that range from a clove-like spiciness or a banana-strawberry aroma to less pleasant smells of rotten eggs and Band-Aids. Diacetyl, a natural emission of yeast, is the most dominant off-flavor. It leaves an artificial popcorn-like buttery or butterscotch taste and a slick coating on the tongue.

As you sip the beer, notice the experience in your mouth. Does the beer feel crisp and cool or full and warm? The warmth comes from the alcohol. Is the beer rich or thin? Is it sweet or sour ... or sweet *and* sour? How does the tingle of carbonation feel? What does that sparkle ignite? Draw from the beer wheel and from your sensory library.

Observe the layers. Beer isn't one ingredient or one taste. What is the glass revealing to you? And how is it revealed? A fruit beer will show off its sweetness and acidity quickly and easily. The bitterness in hops tends to take its time. Pay attention to what rises and falls.

Take a breath and re-slurp that same sip of beer as we do with wine and coffee, again aerating the beer that is already in your mouth. What shows up for you as the beer gets warmer? What components are drawing you into the experience—and what's pushing you away? What lingers in the finish?

Take another sip. Again, observe everything: the sensations in your nose and mouth, the foam lace clinging to the side of the glass, the memories that are stirred. We tend to think of balance in beer as a relationship between bitter (hops) and sweet (grain). It is also an interplay between so much more: between the ingredients and the brewer, between our mouth and the glass, between the microbes in our gut and the microbes in our glass. (The American Gut Project found people who drink alcohol have "greater microbial diversity than those that don't drink alcohol at all," which might mean beer drinkers have healthier digestive systems.[130])

Bring out the snacks. Go beyond corn nuts and pretzels. "Food pairings are one of the best ways to open up our palates to beer," says beer consultant Kelsi Moffitt. Beer and food create an entirely new dynamic: Bitterness can make some foods seem sweeter or more intense, while carbonation clears away the richness of fatty foods.

Reset your palate. Spit out the beer, as needed. Recalibrate your sense of taste with crackers and water, and your sense of smell by sniffing the crook of your elbow.

Rinse and repeat.

BREAD

I. THE TEMPLE

"Forget about the taste," Harpreet Singh, the bright-eyed, bearded attendant at the Golden Temple commanded as he led me through a narrow staircase to a small kitchen where the wheat pudding known as *karah prasad* was being lovingly prepared. "You have come to a place of God."

I had followed my palate (and heart) to Amritsar, India, to trace the taste of *karah prasad* back to one of its most revered sources. In holy settings, *karah* is called *karah prasad*, or *prasad*. It's given to visitors of both Hindu and Sikh temples as a symbol of devotion, loosely akin to a Catholic Communion wafer. As a child and young adult, I had always been drawn in by the flavor of the thick, buttery pudding made of wheat, water, sugar and ghee (clarified butter), so Harpreet's statement was jarring.

Forget taste?

This was my first trip back to India since the death of my grandmother. My nani was, in a word, devout—and highly superstitious and crazy for sweets. I inherited all those things from her. As she recited her morning prayers, I would sit by her side, then gladly accompany her on the temple circuit: visits to both Hindu and Sikh houses of worship, during which time we said our prayers and met pandits who bestowed religious wisdom and, occasionally, astrological insight. They gave blessings in the form of their touch and a small handful of *prasad*.

In all my years of holding out my hands—right nestled inside left—in anticipation of the sweet, buttery treat, I never once realized I was eating a blessing. I knew the offering was holy, but I considered it more of a holy *dessert* than the manifestation of blessing. That was, until I visited the Golden Temple and met the people tasked with making the sacred food.

Officially known as Harmandir Sahib ("Abode of God"), the Temple is located in Punjab, a northwestern state known as the breadbasket of India because it provides over half of the wheat consumed in the country.[1] Three out of every four residents of Punjab are involved in agriculture.[2] It's my ancestral home state; my people are farmers.

Wheat is both a nutritional and spiritual staple. Despite a wave of diets that shun the grain (from Atkins to Paleo)—and a growing number of people who have no evidence of Celiac disease or intolerance but live in fear of gluten—wheat is the staff of life.

According to the Food and Agriculture Organization of the United Nations (FAO), 15 plants provide 90 percent of our global food energy intake, two-thirds of which comes from just three plants: rice, corn and wheat.[3] These three crops are the foundational foods for more than 4 billion people,[4] of which wheat is the most widely cultivated.[5] A grass best known for its grain, it is also used as animal feed, biofuel and straw for housing. Wheat is a crop interwoven with the dawn of civilization and the development of agriculture, starting with the soupy beer-bread that transformed us from hunters and gatherers into farmers.

Wheat lives within the Poaceae family, the fifth-largest family of flowering plants that includes bamboo, lawn grasses, many types of weeds and all cereal grains (wheat, corn, rice, barley, rye, oats, millet and more). It "encompasses a sprawling family tree of species, and myriad varieties within those species."[6]

Durum wheat (*Triticum turgidum*) is the high-quality wheat used for making pasta, breads such as pita and roti, and couscous.[7] It's a natural hybrid of wild Einkorn wheat, which originated in southeastern Turkey,[8] and another wild grass, believed to have originated across the Near East, through what is now Turkey, Syria, Iraq and Iran.[9]

Common wheat (*Triticum aestivum*) makes up most "raised" breads (as opposed to flat ones), and is the grain most widely grown around the world, on every continent except Antarctica.[10] It is a cross between a wild goat grass (what most consider a weed) and a cultivated wheat (most likely the one used to make pasta). "Genetics suggests this cross took place in the Caspian region of Iran; archaeology suggests that this took place about 9,000 years ago."[11]

In the United States, we now eat about 132 pounds of wheat per person per year.[12] From breakfast to dessert, wheat is our staple—cereals and toast, sandwiches and pizzas, pasta and seitan (the meat substitute), plus binders in salad dressings and sausages, and, of course, the beloved foundation of cookies and cakes. Wheat is also the foundation food of faith—our spiritual staple. From the host at Communion to the matzoh at Passover, wheat nourishes our body and soul. It is our daily bread.

At 10:30 P.M., I got out of my warm bed, took a lukewarm shower, donned my *salwar kameez*, an outfit comprising a long tunic and matching pants, and wrapped around my neck my *chunni*, the scarf that would eventually cover my hair. I was tired, but I perked up against the unseasonably chilly night and the chance to observe the preparation of the holiest of foods at the holiest temple for Sikhs in the entire world—a process few have the opportunity to witness.

Sikhism is a faith dedicated to prayer and prayerful actions through a balance of selfless service and self-renunciation.[13] The first words of our sacred book, the Sri Guru Granth Sahib, state: "There is only one God. Truth is his name. He is the creator. He is without fear. He is without hate. He is immortal, without form, beyond birth and death. He is self-illuminated, realized by the kindness of the true Guru."[14] To fully embrace these words—known as the Mul Mantar, the root prayer—is to understand the essence of the faith and the path to knowing God.

But India, the place where Sikhism originated in 1500, is more difficult to comprehend—especially if you've only seen the place from a distance, in print or on screens. It is a contrast of vast wealth and crushing poverty, home to the highest concentration of both billionaires and impoverished people in the world.[15] India boasts sprawls of peaceful, isolated backwaters and suffocating urban jungles. It is beauty and beast in every breath, in every glance, in every touch. And it doesn't necessarily get easier to understand once you've experienced it up close. Mother India is impossible to pin down; she's all over the place.

Yet this raucous and stunning place defines who I am. Despite spending most of my life in the United States, India is the touchstone of my being: many things, all at once. People are constantly touching, pushing, prodding and staring—stares of recognition that, though I look like a Punjabi woman (hooked nose, wheatish complexion), I am not from this place. My gait is too fast; my *salwar kameez* hangs just a little differently on my frame.

India is a sensory overload of touches, smells, sounds, tastes and colors: all of them an intensity I've never experienced anywhere else, all of them doing battle to stand out. The strongest win every time. Except here, in this sacred space. The Golden Temple is an oasis. Out of the many holy sites I have visited in India, it's the only one where—despite 100,000 visitors per day[16]—it feels spacious, and I am at peace.

I had returned to India to understand exactly when and how the wheat pudding I had loved my whole life transformed into a sacred blessing. *Karah prasad* is made not only in temples but also at home. It's a food we eat when we celebrate and mourn that, under certain circumstances, becomes holy.

My work companions—journalist Alok Gupta and documentarian Jason Taylor—and I arrived at the Golden Temple around 11 P.M., close to the time when the *prasad* would be made. We got out of the car and walked silently toward the majestic marble shrine. Shoes are prohibited inside the compound, so we deposited ours with the attendant and stepped into shallow pools of water intended to rinse away any residual dirt, then onto the cold marble floor of the Temple grounds. As soon as we caught sight of the gilded structure that houses the Sri Guru Granth Sahib, the holy text that is considered the final incarnation of Sikh spiritual teachers, I pressed my palms together in reverence. The reflection of the Temple shimmered in the water that surrounded it. Its beauty made my throat tighten. Harmandir Sahib was awe-inspiring but welcoming—a home of God.

This deep-rooted connection has happened to me only occasionally: in Cambodia's Angkor Wat, in select cathedrals in Rome, and in the dim basement of a community center in Alpine, New Jersey, where a blind rabbi named Dennis Shulman convenes his morning minyans.

I found God in a folding chair in New Jersey—and I found God here, in the gleaming golden temple resting in the middle of a lake of *amrit*, holy water.

Alok dropped to his knees and touched his forehead to the smooth tiles. This gesture, practiced across many faiths, is one of surrender. I was a bit more squeamish and engaged in a modified form of prostration, touching my hand to the tiles, then to my forehead and heart. I felt quiet all around me as my shoulders released, as *I* surrendered.

Alok greeted the officials who then led the three of us into a narrow circular passage and up the stairs to the place where wheat, sugar, butter and water become *prasad*.

Diane Ackerman writes in A *Natural History of the Senses*, "Taste is an intimate sense. We can't taste things at a distance."[17] But for me, the most intimate of sensations is touch, the sensation felt most acutely in India. "Touch," she writes, "is the first of the senses to develop in the human infant, and it remains perhaps the most emotionally central throughout our lives."[18] It is a constant: We can close our eyes, pinch our noses shut or smoosh in earplugs, but "there's no way to turn off touch," explains neuroscientist David Linden.[19] "It's always there, and because it's always on, we take it for granted."[20]

Our skin perceives different types of stimuli, including pressure, temperature, pain and vibration, all of which are received through touch receptors in the skin.[21] These nerve cells fall into two groups: mechanoreceptors, which transmit information on touch, pressure, vibration and tension felt on the skin (such as pushing or pulling)—and thermoreceptors, which detect heat and cold.[22] Although touch receptors are scattered all over our bodies, they are concentrated in certain places: on our lips and inside our mouths, and on the tips of our fingers and the soles of our feet (where more than 3,000 receptors are clustered).[23]

As we walked toward the kitchen, the first thing that hit me was the ghee—not just the scent of the clarified butter, but the texture. The stairs leading up to the cooking area were greasy, the walls slick to the touch; my bare feet slid slightly with each step. The entryway was thick with the sensation of *prasad*. As we walked into the kitchen, the three

Amritdhari halwais (makers of the holy food) did a double take: This was not a place for women. I tucked my hair into my headscarf and pulled it a little tighter.

Throughout my life, I have struggled to understand these gender norms and, in many instances, to defy them. And I have struggled, in this book, to reconcile the fact that, while most of the book has been built on the sweat and hopes of women, men receive the most attention and reward. Men dominate in research science, including conservation—and they have assumed the lead in wine and chocolate making and in coffee roasting and beer brewing. Although women make up almost half of the agricultural sector in places like India, they have limited formal education and access to land.[24] Men dominate in agricultural institutions, and men dominate here, at the Golden Temple.

Yet women remain indispensable.

In celebration of my 40th birthday, food activist and chef Defne Koryürek and I cut into apple cake as we sat at our favorite restaurant in Istanbul. I was explaining to her that, when I was coming of age in college (as a hairy-legged feminist), I decided I wouldn't do the things my mother had done. I didn't want to cook or clean. I wanted to be the kind of woman who had the financial means to pay someone else to do those things, so I could dedicate my time to other (presumably nobler) tasks.

She cocked her head and said, "You know you were wrong, right?" I did. But I didn't understand the depth of that wrongness until she added, "Baking bread is the most revolutionary thing you can do."

I have worked hard to live a life of integrity, one in which words and actions align. That's why I sneak my own healthy snacks (ranging from organic smoothies to seaweed strips) into movie theaters and carry a reusable water bottle onto planes, suffering the withering glances of flight attendants who want to pour my water into a plastic cup and move on to the next passenger: "Can you fill up this bottle, please?" These acts seem small, almost inconsequential, but they are built upon the stubborn belief that how we spend our days is how we spend our lives. I want to spend mine celebrating the things I love, honoring the people and circumstances that ensure my little joys will endure. That's why I grind those coffee beans and pay more for good chocolate. I won't singlehandedly transform the commodities market, but I will take the opportunities this life provides and use them the best I can. If I am going to spend $6 on a loaf of bread or $12 on a bottle of wine, I want them to be the ones that reflect and support what I value.

This isn't to say I don't occasionally relent and take the plastic cup, eat the Twix or drink the cheap wine. Every choice I've made is a response to my life and to the moment in which I find myself. It's the same with taste: We carry our stories to every decision on what and how we eat. I didn't want to cook because I didn't want to be like my mother. But she has evolved into a woman who hates cooking, which means I am *just* like her. This, I now realize, is a wonderful thing.

I crave the care embedded in well-cooked meals, and I aspire to, someday, extend more of that care to myself, through my own hands, as my own tiny revolution. But today, I am content to fully support the women and men who tend those flames, who transform the lives of those around them through their food.

One of the defining characteristics of any Sikh temple, anywhere in the world, is *langar*, the free and open community kitchen mandated by the leaders of the faith where all are fed. Anyone who enters a Sikh temple—rich, poor, young, old, Muslim, Christian, man, woman—will be given a meal. This is accomplished only through the efforts of volunteers who spend time preparing food as a manifestation of *sewa*, selfless service. "By serving food to others," writes Alice Peck in *Bread, Body, Spirit: Finding the Sacred in Food*, "we nurture them, to be sure. [But] in so doing, we are simultaneously tending to the Sacred within ourselves and our world."[25]

Kulwant Singh, a handsome 48-year-old man with a straggly beard and slight wheat belly, explained, "You can't have peace of mind when you are hungry. You can't be nourished spiritually if you are hungry." *Langar* feeds both these hungers.

In the Temple kitchen, women (working alongside men) take the lead in food preparation. Within the cultural context of India, this is radical: first, because a man's place is rarely in the kitchen (unless we're addressing house workers and the inequity of caste divisions, an entirely different conversation), and second, because, if a man *is* in the kitchen, he's probably not working alongside a woman—and definitely not taking direction from her. The reconfigured hierarchy of *langar* is a reflection of Sikhism's emphasis on gender equality, the Golden Temple officials insist.

The reason I could not stand on the platform within the kitchen where the *karah prasad* is made, I was told, was because—in *that* kitchen—they were actually trying to subvert the dominant model of women cooking for men: "Everyone queues to receive *prasad*—rich, poor, famous, infamous, disabled or abled—everyone is equal. *Karah prasad* is the taste of equality." I wasn't fully convinced, but I enthusiastically agreed with them that women are the backbone of both food and agriculture. Without them—without us—we fail.

At 11:45 P.M., nearly an hour into our visit, Kulwant summoned the aide to bring us tea. It was time for Kulwant and his colleagues Bhupinder Singh and Jagdish Singh to make *prasad*. The six *halwais* work in two shifts (from 3 P.M. to 9 A.M.), ensuring every person who makes this pilgrimage to the Golden Temple will receive the edible blessing.

The men looked gallant in tightly wrapped blue and black turbans, orange scarves, white tunics and pants with bright cummerbunds, into which were tucked ceremonial swords known as *kirpans* (one of five articles of faith worn by baptized Sikhs).[26] Although not all are related, all Sikh men bear the surname Singh (meaning "lion"), bestowed with the intention of eliminating discrimination "based on 'family name' (which denotes a specific caste) and [reinforcing] that all humans are sovereign and equal under God."[27]

The men began to hoist 15-kilogram (roughly 33-pound) tins of ghee, sugar and wheat onto the platform where massive cooking vessels (*karais*) were set on top of large gas grills the length of bathtubs. Jason leaned in and said, "It's like the kitchen of a giant." Exactly. The metal tins, the stirring spoons, the *karais*—everything was supersized; it was *Guru ka rasoi*, the kitchen of God, one that produced enough *prasad* for each of the 100,000 daily visitors to consume on-site and carry home.

The chai arrived in small metal bowls so hot we had to hold them by their edges. It was steamy, milky, fragrant with anise—and much needed. It was now close to midnight. With a loud whoosh, *halwais* Kulwant and Bhupinder fired up the gas burners, then banged open the giant tins and began to dissolve the sugar in water. A few minutes later, Jagdish sloshed the ghee into another vessel, followed—in equal proportion—with a tin of coarse wheat flour. He began to stir and stir. The preparation of this portion of *prasad* would use one tin each of wheat, sugar and ghee, plus 45 liters (or nearly 12 gallons) of water.

Between the igniting, banging and stirring, talking became prohibitive. I was grateful, as my thoughts had started to drift to my grandmother, the woman who had most influenced my perceptions of food. I imagined her standing beside me, her plump hand in mine, taking in the grandeur of the Golden Temple and contrasting it with the familiarity—the universality—of slaving over a hot stove. It was a moment of such deep reverence and knowing. Alok broke the loud silence to say, "You can feel it." Yes, I could.

The room got warmer and warmer, and the air became heady with the smell of propane gas, roasting wheat, Jason's sandalwood oil and the rising smell of sweat from the laboring *halwais*, who had removed their long-sleeved shirts and were working in undershirts that were translucent with perspiration. Temperatures inside this room are known to climb to a smoldering nearly 47 degrees Celsius (116 degrees Fahrenheit).

As the mixture of wheat and ghee thickened, it became increasingly difficult to stir. The men were getting tired. They took turns stirring the *karah*, mopping sweat from their faces with their bright orange scarves as they repeatedly chanted "Waheguru" (best translated as "God is great") under their breath. It was the second preparation of their shift. Everything was slowing down.

Then, as if on cue, the gated elevator clanged open and three more men walked into the kitchen. They stepped onto the platform and helped the *halwais* lift the heavy vessels of sugar water into the buttery wheat. They took over stirring as steam billowed from the pots, all the while chanting "Waheguru." As Kulwant stepped off the platform to take a break, I asked when he knew the *karah prasad* was ready. "You know by the texture," he said. Bhupinder somehow overheard my question and shouted from the platform, "You know as the color gradually changes." With that, he lifted up a thick spoonful of pudding that was now golden brown, like the color of our skin. "This is all a stepping stone to know the Guru, to know God," Jagdish said. "As I make this *karah prasad*, I am transforming myself for God and the community."

An attendant brought more chai, this time even sweeter, in cups the size of soup bowls. The *halwais* gulped the steamy liquid down with one hand. I held my cup with two and sipped slowly as I slid my toes back and forth on the greasy floor. These walls and floors—every surface—held years of accumulated ghee, dabbed-away sweat, whispered prayers; moments when water, two plants and the offerings of an animal became something sacred. "The more engaged we are with the Guru," Bhupinder said, "the better the outcome."

The next morning, Alok, Jason and I returned to the Golden Temple, a place that felt only slightly less spacious in the light of day when the majority of worshippers were in attendance. We removed our shoes, dipped our feet into the small pool of water, and paid our respects before being ushered up to a small room where a group of a dozen Golden Temple workers were gathered.

"The taste doesn't matter," repeated Harpreet Singh, the enthusiastic 34-year-old assistant with a saffron turban and slightly greying beard. "And that's why it tastes better. What you taste is not the driving force, it's what you *feel*—passion, *sewa* [service] and devotion." I still had questions: Was this—the gratitude of service—what made *karah prasad* holy? I traced the ingredients all the way back to their source—to the plants and animals, and the people who raised them—to try to understand where sanctity begins.

Karah prasad is made up of equal parts sugar, wheat and ghee to reflect the equality of all who are served; everyone receives the same amount, served from the same dish. Devotees often donate the ingredients but, because that supply is inconsistent, officials make public solicitations for sugar. The clarified butter (ghee) comes from Verka, the state of Punjab's

Cooperative Milk Producers' Federation, more popularly known as Milkfed Punjab. In India, ghee can refer to vegetable oil, sunflower oil or any other fat-based oil. "But when one mentions *desi* ghee," Manjit Singh Brar, managing director of Verka, explained, "it means clarified butter made from cow's milk." One hundred percent *desi* ghee is considered a mark of purity and pride, albeit for reasons that have changed over time. "Actual *desi* ghee used to be made from only native cows," he said, "but now the milk comes from various cow breeds."

The top milk producer in the world, India is responsible for 18 percent of global production, nearly one out of every five glasses the world drinks. The second-largest producer is the United States, with 12 percent, while Brazil and China tie for third, each representing 5 percent of production.[28]

These cows aren't worshipped, but they are respected. As animals that provide milk, fuel (from manure) and labor, cows have a special place within Indian culture as a symbol of Mother Earth and all the gifts she selflessly bestows.[29] Yet, as heralded as cows are supposed to be, the 34 breeds of native cows that have thus far been identified have become a lot less common.[30] Just as crops have been replaced with high-yielding varieties, so have animals.

With the Westernization of agricultural research and education, food and policy analyst Devinder Sharma explains, scientists "considered it worthless to work on the native breeds."[31] Cattle improvement efforts, instead, focused on crossbreeding foreign breeds (Jersey and Holstein-Friesian cows) with local ones to improve domestic milk production. The imported Jersey purebreds produce an average of 800 to 1,300 gallons of milk in a lactation year, but their offspring (which are crossbreeds) only give about 660 to 800 gallons of milk. Under local environmental conditions, Devinder contends, India's three indigenous breeds—the Gir, Kankrej and Ongole—give more milk than Jerseys and Holstein-Friesians. Productivity isn't about just the cow but the context in which it's placed.

The tragic irony, Devinder says, is that "Brazil has ... emerged as the world's biggest supplier of improved cattle embryos and semen of Indian origin—now rated amongst the best dairy breeds in the world." This is due to a combination of factors, explains Paul Boettcher, animal production officer in the Animal Production and Health Division at FAO. "For better or worse, biodiversity laws strictly regulate the exportation of native cattle from India which also contributes to Brazil being the main exporter." If policymakers and "Indian dairy and animal scientists had not ignored the domestic cattle breeds," Devinder adds, "the fate of the Indian cows would have been much different—these holy cows would have then been truly revered."[32]

And they would have transformed the entire world of dairy production. The Westernization to which Devinder refers has reshaped not only dairy production and cattle breeding, but every component of Indian—and global—agriculture. Including wheat.

The *atta* (wheat flour) for the Golden Temple comes from donations, plus 300 acres of farmland owned by the Temple. I reasoned that, since the Temple (like all temples) was resolute about the use of *desi* ghee, the wheat and sugar—equal components of *karah prasad*—would also be scrutinized for purity. That's why I was stunned when the wheat farmers told me their crops were full of *dawai*: poison. *Dawai* is the Punjabi (and Hindi) word for "medicine," but in this context, *dawai* means "poison."

In order to understand how wheat farmers came to describe their crop as poison, we have to rewind to the late 1960s, the dawn of what's known as the Green Revolution. If you were around at the time, you might remember television commercials showing image after image of starving Indian children with bloated bellies and torn clothing. I wasn't born until 1970, but those messages still dominated television airwaves when I was growing up in North Carolina. Even my own mother told me to "think of the starving children in India" as she urged (forced) me to finish all the food on my plate.

India's famine was real—and devastating, the tragic result of political, socioeconomic and agricultural factors. It had its roots in the British occupation of India and the Bengal Famine of 1943, when three million people starved to death, not only because of the loss of rice crops (due to disease, a cyclone and three tidal waves), but as a result of poverty, inequitable distribution of remaining food grains, the hoarding by the wealthy of whatever rice stocks were available, and the crushing governance of British prime minister Winston Churchill. Author and politician Shashi Tharoor documented in his article "The Ugly Briton" that Churchill, "as part of the Western war effort, ordered the diversion of food from starving Indians to already well-supplied British soldiers and stockpiles in Britain and elsewhere in Europe, including Greece and Yugoslavia. ... Churchill's only response to a telegram from the government in Delhi about people perishing in the famine was to ask why Gandhi hadn't died yet."[33]

The suffering didn't end there. "Back-to-back droughts in India during the mid-1960s made the already precarious situation worse, and a 1967 report of the U.S. president's Science Advisory Committee concluded 'the scale, severity and duration of the world food problem are so great that a massive, long-range, innovative effort unprecedented in human history will be required to master it.'"[34]

The initiative was led by the Ford Foundation, the Rockefeller Foundation and the U.S. government, working together to develop a global agricultural research system. This became the foundation for the Green Revolution, a term coined by USAID administrator William Gaud to describe "the vast potential of agricultural technology."[35] The system was built on several components: 1) the expansion of agricultural land; 2) the introduction of improved, or high-yielding varieties (HYVs) of, seeds, which grew into resilient dwarf plants that wouldn't fall over (unlike their predecessors); 3) the introduction and use of synthetic fertilizers and pesticides to amp up and protect yields; and 4) irrigation to artificially create an additional crop season that didn't depend on monsoon rains.[36] In short, the Green Revolution was the industrialization of agriculture. It came with all the trappings of what the West stood for: efficiency and modernity.

The intentions were noble—and the Green Revolution was, initially, incredibly successful. It helped increase food production, fed and empowered people, and established India as one of the world's largest agricultural producers: "Instead of widespread famine, cereal and calorie availability per person increased by nearly 30 percent, and wheat and rice became cheaper."[37] Over the next two decades, localized varieties of wheat (known as landraces) were replaced with improved, high-input varieties. Nearly three-fourths of the land dedicated to growing wheat was sown with seeds that, at the time, helped make India not only self-sufficient in grain production but an *exporter* of food.[38]

At the forefront of the charge to increase India's self-sufficiency was Mankombu Sambasivan (M. S.) Swaminathan, a scientist and plant geneticist widely considered the "Father of the Green Revolution." His response to hunger in India was both personal and political: "At the time all our young people, myself included, were involved in the freedom struggle [from British imperial rule], which Mohandas Gandhi had intensified, and I decided I would use agricultural research in order to help poor farmers produce more."[39]

Throughout Punjab, Swaminathan planted seeds developed by geneticist and plant pathologist Norman Borlaug, and worked with Indian geneticists and local farmers to crossbreed Mexican wheat with local and Japanese strains. Together, they created a semi-dwarf wheat that yielded significantly more grain than traditional types.[40] "By combining scientific research with local knowledge, [and] enabling policies and communication technology, wheat production went up from 10 million tonnes in 1964 to 17 million tonnes in 1968. Production now [in 2015] exceeds 97 million tonnes."[41]

Punjab was India's, and the Green Revolution's, greatest Cinderella story: the begging bowl became the breadbasket.[42]

This transformation was so great that Borlaug was awarded the Nobel Peace Prize in 1970 for his work developing HYVs of wheat. Like Swaminathan, Borlaug made clear in his Nobel speech that his work originated from a deep commitment to his values ("Food is the moral right of all who are born") and social justice ("If you desire peace, cultivate justice, but at the same time cultivate the fields to produce more bread; otherwise there will be no peace"). It revealed how deeply he—like M. S. Swaminathan—cared.

"Today," Borlaug said in his acceptance speech, "we should be far wiser; with the help of our Gods and our science, we must not only increase our food supplies but also insure them against biological and physical catastrophes by international efforts to provide international granaries of reserve food for use in case of need. ... Since the urbanites have lost their contact with the soil, they take food for granted and fail to appreciate the tremendous efficiency of their farmers and ranchers, who, although constituting only 5 percent of the labor force in a country such as the United States, produce more than enough food for their nation."[43]

Those words were spoken in 1970, but still hold true today. Except now, less than 2 percent of the U.S. labor force works in agriculture.[44] And while roughly 40 percent of our food ends up in landfills, 50 million Americans go hungry.[45] The United States is the world's largest exporter of wheat but not its greatest producer.[46] That title goes to the European Union, China and India.[47] Today, decades after the famine that ravaged India, our televisions broadcast messages encouraging viewers to help feed hungry children ... in America.

There is nothing noble about being hungry or making agriculture harder than it needs to be; however, the transformations of the Green Revolution didn't come without trade-offs. The improved seeds were created from a small collection of HYVs developed through two decades of research and breeding of dwarf varieties of wheat in Mexico (and, subsequently, India). They were part of a broader set of cultivation practices intended to, as Borlaug said, "cure [the] ills of a stagnant, traditional agriculture."[48]

The traits held in those seeds—disease and pest resistance, drought and salinity tolerance—have been bred into new varieties of wheat and hold the keys to the future of one of the world's most important crops. But the focus on select seeds dramatically eroded the existing gene pool. Removing traditional varieties from fields not only caused genetic erosion, it minimized the importance of farmers' experience and the knowledge that came from generations of working the land, growing crops and understanding, intimately, the specificities of place. It was cultural erosion, too.

While seeds saved in *ex situ* collections are incredibly precious, *in situ* conservation is vital. It ensures crops—and the people who grow them—can respond to changing social needs and environmental conditions. But, unfortunately, that's exactly what the Green Revolution discouraged. Farmers were pushed to abandon traditional ("backward") cultivation methods and embrace new seeds over old, chemical fertilizers over cow manure, tractors over bulls, and clean monocropped fields over the "messy" intercropped fields they had traditionally sown.[49] They were told to give up what they already had (the seeds, the animals) and increase their dependence on what they'd have to buy (new seeds, plus all the inputs required for them to flourish).

It was the modern way—but it was also a mistake.

While roughly one-fourth of food crops are preserved in genebanks, Penn State geographer Karl Zimmerer and his colleagues learned, through their assessment of global census data, that "up to 75 percent of the seeds needed to produce the world's diverse food crops are held by small farmers" working on plots of less than 7 acres. Most of these farmers are women who preserve diversity through "networks of seed and knowledge exchanges."[50] They grow crops that flourish in small areas and nourish local palates, ones that may not ever make it into limited *ex situ* collections (which, however diverse, can only hold so much). These women are the ones securing the foundation of food sovereignty: the right to decide what to grow and how to grow it—and the right to ultimately decide what to eat.[51] Women feed the world.

In a story titled "India's Farming 'Revolution' Heading for Collapse," investigative reporter Daniel Zwerdling explains, "When farmers switched from growing a variety of traditional crops to high-yield wheat and rice, they also had to make other changes. There wasn't enough rainwater to grow thirsty 'miracle' seeds, so farmers had to start irrigating with groundwater. The high-yield crops gobble up nutrients like nitrogen, phosphorous, iron and manganese. ... [As a result,] the farmers say they must use three times as much fertilizer as they used to, to produce the same amount of crops."[52]

FAO director-general José Graziano da Silva has admitted the Green Revolution "showed its limits." The focus on a few crops reduced diversity, and the intensive use of water and chemical inputs compromised the environment. "It is rightfully credited to have saved over one billion people from starvation in the last decades," he said. "But it wasn't able to end hunger. Increasing production with the same input-intensive approach take[s] too heavy a toll on the environment."[53]

No matter where you go in India, you are reminded that hunger has not ended and that industrialized agriculture's toll has been too heavy—especially in Punjab. The cost is reflected in water poisoned by run-off from synthetic nitrate fertilizers, soils contaminated with dangerously high levels of heavy metals from pesticides and on-farm vehicles, and children suffering from medical ailments ranging from brain damage to cancer.[54]

Harpreet, the attendant at the Golden Temple, explained this vicious cycle was an attempt to do the right thing: "It was a matter of survival," he said. "It was the government who brought the dam, the agronomists who taught us to use pesticides and insecticides. Everyone knows chemical farming is dangerous, but, since it has become part of farming in Punjab, no one seems to care if it's going into our belly or into the Temple. The only wheat that is available has pesticide. Even if you have money, you have to buy what's in the market."

This contradiction didn't make sense to me. Officials were obsessed with the purity of the ghee, but the wheat—the crop that defines Punjab, home to the largest Sikh population in the world—could be anything. Including *dawai*.

Although new policies put into place by Temple officials indicate this might change,[55] "records at the Health Department of the Government of Punjab show 34,430 people died due to cancer in the last five years—20 deaths each day."[56] Add to this, the shockingly high incidences of suicide among farmers: According to data from India's National Crime Records Bureau, more than 250,000 Indian farmers have taken their own lives since 1995. "That is around 46 farmers' suicides each day, on average. Or nearly one every half-hour since 2001."[57] Most of these farmers killed themselves with the poison that was most readily accessible: pesticide.[58] And this isn't India-specific. The suicide rate for male farmers in the United States is nearly double that of the general population.[59]

The cost has been too high.

As Alok, Jason and I headed toward villages outside of Amritsar to meet wheat farmers who had experienced the effects of the Green Revolution firsthand, I secretly hoped Harpreet was wrong. In fact, I *desperately* wanted him to be wrong, because the implications of him being right were simply too painful—for so many people.

Jashant Singh and his son, Jagdish (no relation to the *halwai* at the Golden Temple), were two of the handful of farmers Alok had arranged for us to meet. Jashant greeted us with the traditional Sikh greeting "Sat Sri Akal" ("God is truth") and a warm invitation to sit down. More than any other farmer we met, Jashant looked like a relative. We are all Punjabi—we all have the same nose and complexion—but Jashant stood tall, smiled wide and reminded me of my maternal uncles. His 20-year-old son—in jeans and a polo shirt, and with, for most of the conversation, thumbs on his iPhone—reminded me of my wealthier New Delhi cousins.

Over cups of chai and wheat Marie biscuits, Jashant explained, "The *dawai* and fertilizers were forced on us. We're trying to get out of the clutches of using chemicals on our farmland, but, without these things, the seeds perform worse than traditional ones." This refrain of "more chemicals equals higher yields" echoes throughout industrialized agriculture. The chemical dependence experienced by wheat farmers and other farmers can be compared to any other addiction: It feels good in the moment, but, in order to achieve the same effects, you end up using more and more of the drug.

There appeared to be no end in sight to this dependence until I asked Jashant if he and his friends used the *dawai* on the crops they grew in their home gardens for personal consumption. Jashant shook his head no.

That's where I lost it. I didn't mean to—I didn't want to—but my voice started to shake and I began to tear up. "The commercial wheat you grow is what feeds me," I said. "I am a Punjabi girl. I'm like your daughter, and you are feeding me poison, poison that stays all around you." Poison, I was too upset to say, that affects millions of people, whether they are aware of it or not, in India and around the world.

I was hoping for an epiphany, some reflection that this decision impacts people, the soil, the water. All Jashant said was "Becoming a slave is very easy, but getting free takes time. Reclaiming this takes time."

Jashant wasn't only referring to growing methods. Industrialization starts in the seed, the building blocks of food. Although public breeding programs for wheat are still strong,[60] and new wheat varieties are largely developed through standard plant breeding, many of our food staples are increasingly grown from hybrid or genetically engineered seeds, not traditional ones.

Traditional seeds can be saved and used for subsequent plantings. They are what Utah farmer and breeder Joseph Lofthouse describes as "genetically diverse [and] promiscuously pollinated,"[61] meaning open-pollinated seeds with no restriction on the flow of pollen between individual plants.[62] Older varieties of these plants are often referred to as heirlooms. Heirlooms differ from hybrids, which can occur naturally over time but are often the result of a particular, carefully planned cross between similarly related species. Pluots, for example, are crosses of apricots and plums, while seedless watermelons are crosses of parent watermelon plants with differing numbers of chromosomes that render sterile (seedless) offspring. Modern carrots are a cross between Dutch carrots and wild carrots that originated in Afghanistan some 5,000 years ago.[63]

Traditional seeds also differ from genetically engineered, or transgenic, seeds that are widely—but incorrectly—referred to as "genetically modified organisms," or GMOs. They are seeds that are created "when one or two genes with the desired traits from *any* living organism are transferred directly into the plant's genome."[64] There is some evidence that this has occurred naturally,[65] but most transgenic changes occur in laboratories.

When commercialized, both hybrids and genetically engineered seeds have to be repurchased every year because 1) they won't render consistent results if saved and used in subsequent plantings, and/or 2) they are protected as the "intellectual property" of seed companies.

The question of how much of what Mother Nature has given us—what's evolved over centuries and is now allowed to be patented—is debatable. But genetically engineered (transgenic) seeds have faced additional scrutiny for other reasons, most notably for the kinds of traits bred into the crops: Most of this seed has been commercialized and developed for herbicide and insect resistance by large-scale agribusinesses, organizations that prioritize their return on investment over a seed's nutritional value or social impact.

While scientists, health advocates and the public continue to debate the long-term health and environmental impacts of transgenic crops, there is near consensus in one area. "What is clear," explains Bioversity International's principal scientist Mauricio Bellon, "is that the business model of a seed company goes against traditional breeding models." Monsanto and Syngenta, two of the top global producers of genetically engineered seeds, are also the world's top sellers of all commercialized vegetable seeds.[66] These concerns are not limited to biotech seeds, but *all* seeds.

According to USDA data from 2007, Monsanto and Syngenta, along with two other companies, control more than 70 percent of the *entire* proprietary seed market. The top eight companies control 94 percent of the commercial market and, the USDA report adds, "these shares have most likely continued to rise."[67] Those who control seeds control food.

As I wiped away tears with the corner of my *chunni* and tried to compose myself, Jagdish posed a question: "Excuse me, ma'am, do you work in an office?"

I smiled. "Most of the time," I said, grateful for the distraction.

"You see, ma'am, I want to work in an office, too. I want to work inside, with air-con. I don't want to grow this poisoned wheat."

I then understood. Jagdish didn't want to grow any wheat *at all*. His father was working as hard as he could to ensure Jagdish would have an opportunity for a life closer to mine.

Every decision has trade-offs. In India and around the world, millions of farmers are working to transform a system that initially fed them but has, ultimately, bound them. They are trying to marry agricultural wisdom developed over millennia with modern technologies that can make their lives easier and more profitable—without becoming dependent on a single system or beholden to corporations' claims of intellectual property. These farmers want the option to choose diverse seeds, to choose work in an office cubicle over work in a field—to regain sovereignty over their food and their lives.

They want the freedom to choose.

But the sad truth is, Jagdish *will* likely end up in an office, not necessarily out of choice. Daniel Zwerdling reports the water tables in Punjab keep dropping and, if agricultural methods don't change, Punjab "could trigger a modern Dust Bowl." Wells that had been drilled 10 feet deep have now been deepened to more than 200 feet. "The heartland of India's agriculture," he concludes, "could be barren in 10 to 15 years."[68]

Later that night, Alok, Jason and I went to dinner at the home of my extended relatives, distant relations who, nevertheless, treated all three of us like members of their immediate family. The meal was simple and typically Punjabi: *rajmah* and *chawal* (kidney beans and rice) and *bhartha* and *roti* (roasted eggplant and flatbread). Jason is British, Alok is from a different part of India; the meal was a reflection of *my* heritage.

Throughout the meal, I kept thinking about queries posed by ethnobotanist and conservationist Gary Nabhan in his seminal book *Coming Home to Eat: The Pleasures and Politics of Local Food.* "What do we want to be made of?" he asked. "And what on earth do we ultimately want to taste like?"[69]

These questions start as seeds: What we do to our ecosystem—what we do to our food—is what we do to ourselves. When we talk about depletion in soil and water, we're also talking about something being depleted in *us.* When we refer to monocropping and monodiets, that monotony is a reflection of how we live our lives. If the stories of seeds are the stories of us, we have to ask ourselves, "What seeds are we planting—and how do we nourish the seeds we want to grow?"

II. THE BLESSING

Zoologist Charles Godfray, the president of the British Ecological Society, once said, "If we fail on food, we fail on everything."[70] But we have too much at stake—too much taste, too much memory, too much of our story—to fail. We are in this together, dependent on farmers, breeders and stewards of biodiversity from all over the world who sustain us. With every bite, we have shaped and will shape what and how we eat. This is *in vivo* conservation; it is what happens in our living and doing, by consuming in a way that reflects and contains our cares.

I understand why Harpreet Singh asked me to "forget taste" while eating *karah prasad*, but taste—flavor—matters because it can forge a different kind of change. It's what inspired Dan Barber, executive chef of New York City's Blue Hill restaurant, to take what he calls a "whole-farm approach" to meals: "My aha moment," he explained to me, "happened one day in the kitchen, as I was looking over a bag of all-purpose flour. We use a ton of flour at Blue Hill. It's probably the most ubiquitous ingredient in our restaurant. But I realized that day I knew *nothing* about it. I didn't know where it came from or how it was grown—I only knew it had absolutely no flavor, and it was in *everything*. There I was, running a farm-to-table restaurant, serving only the best organic vegetables and the most humane free-range meats, but when it came to wheat, I was cooking and baking with what is essentially rotten produce. So I decided I wanted to get my hands on some delicious local flour—flour from wheat with a story."

In focusing on what Dan calls "the all-stars of the farmers' market," such as heirloom tomatoes, we've overlooked "a whole class of humbler crops that are required to produce the most delicious food." He continued, "Diversifying our diet to include more local grains and legumes is a delicious first step to improving our food system."[71] Now he looks not only at the crops he desires (what he describes as a "grocery-aisle mentality"), but at the potential of everything grown on the farm, including the cover crops and rotational crops that are part of that same system—one that starts with the seed.

"A seed doesn't just contain biological markers, it also comes with social and political markers, and is a holder of memory, heritage, and tradition," writes GBNSP Varma in his piece on Indian ecologist and seed hero Debal Deb, a man who—with the help of local farmers—has sustained rice diversity in eastern India.[72] On just over 2 acres of leased land, Debal grows 940 varieties of indigenous rice that he has collected over nearly 20 years. And they have already proven to be invaluable: "When cyclone Aila devastated the Bengal Delta in 2009, [Debal's] seed bank distributed four traditional salt-tolerant varieties to farmers in the Sundarban Islands. The farmers who sowed those seeds were the only ones who reaped a harvest the next season."[73] Varma summarizes, "Seeds connect us to the continuum of life." They are the beginning and end, then the beginning all over again.

Debal's dedication—and the painstaking efforts of countless others—makes me think of the African woman, never identified by name, who tucked grains of rice into her hair before she was forced onto a slave ship, bringing to the southern United States not only the rice she would cultivate but also knowledge of how to grow the crop—knowledge that enabled her descendants to survive on plantations. Also, of Baba Budan tying seven coffee seeds to his belly and breaking the monopoly on coffee cultivation. And of Nicolai Vavilov, arguably the world's greatest plant explorer, traveling to 64 countries on five continents to collect seeds as he developed his theory on the eight centers of origin of cultivated plants. His work became the foundation of all future work on crop diversity.[74]

Vavilov began this work in 1916 and worked tirelessly until his arrest in 1940 in the Soviet Union. "After over a year and a half of eating frozen cabbage and moldy flour," explains biodiversity expert Gary Nabhan, "he died of starvation. The man who taught us the most about where our food comes from and who tried for over 50 years to end famine in the world died of starvation in the Soviet gulag."[75]

One year later, Vavilov's colleagues in Leningrad (now St. Petersburg) suffered a similar fate. During the Siege of Leningrad, with Nazis bombarding the city, a group of scientists barricaded themselves in the Pavlovsk Experiment Station "to safeguard the 400,000 seeds, roots, and fruits stored in [what was then] the world's largest seed bank."[76] After moving the plant materials into the basement, they worked in shifts to continuously watch over the seeds.

The winter of 1941 was punishing: Food supplies to Leningrad were cut off and it was *freezing*. To survive, people boiled leather belts and purses into jelly. Some ate cats and dogs, while others resorted to cannibalism.[77] Among the thousands who died of starvation were scientists and staff at Pavlovsk. "Some thirty scientists and staff died," writes agriculturalist Cary Fowler, the former head of the Global Crop Diversity Trust. "The curator of the rice collection died surrounded by bags of rice. [Scientists] Kameraz and Voskrensenskaia succumbed, protecting their potatoes in the cellar to the very end."[78]

The collection was too precious to eat.

More recently, conservationist Ahmed Amri and his colleagues at the International Center for Agricultural Research in Dry Areas (ICARDA) genebank in Aleppo, Syria—an area that is part of the Fertile Crescent where wheat was first domesticated—smuggled nearly 150,000 seed samples out of the region before the civil war reached the northern part of the country where the Center's headquarters were located. This included ancient varieties of wheat and one of the biggest collections of barleys and lentils in the world.[79] Now, nearly 100 percent of the seeds are safely stored in backup collections.

Seeds are our legacy—our food, our work and our past, present and future. Protecting this tradition, while championing genetic and cultural diversity, is also the life mission of baker and conservationist Eli Rogosa, a seed advocate and founder of the Heritage Grain Conservancy, a 12-acre farm in Colrain, Massachusetts, that sustains a rich variety of ancient grains and other food crops.

"Modern wheat," Eli explains, "the most widely cultivated crop on earth, is bred by industrial breeders for uniformity and high yield in favorable environments with little regard to the needs of traditional and organic farmers with low-input field conditions, or markets that value taste, nutrition and local cuisine."[80] Eli grows underutilized crops (such as emmer wheat) and landraces, what she describes as "the living embodiment of a plant population's evolutionary and adaptive history."[81]

These grains are commonly known as heirlooms, a term that's a bit fuzzy but is generally defined as older, open-pollinated varieties with a history of having been saved and passed on.[82] They are the kinds of seeds Debal, Indian wheat farmers and countless conservationists have saved—and the kinds of plants Peter, Alberto and Vicente, and Tebeje and Tadesse have grown.

Wheat breeder and plant geneticist David Marshall clarifies, "Landraces are varieties that were selected by our ancient progenitors, who were amongst the first people who started agriculture. Those varieties have been passed down over the eons and are the original selected lines from nature which different people found to be productive to provide food." However, Eli stresses, productivity wasn't just about yield and a handful of agronomic traits, but the role of these plants—in this case, the centrality of wheat—within a culture, about what was created in community.

These characteristics aren't uniform. One culture may value a plant for its leaves; another might eat the flowers. In the part of Ethiopia where David did his fieldwork, some barley and wheat landraces are valued for being tall, a trait that has been bred out of modern wheat because the stalks fall over and can't be mechanically harvested: "A lot of people use that straw for purposes of housing, bedding, whatever it may be," he says. "Today, most of the landraces are only grown in very remote areas of the world. And the rest of them are in seed depositories. They can be used for different purposes, and that's why we need to have them in those collections—so we can go back and see, well, what did they have going for them? What sort of things can we take from them to put into a modern context and be able to make good-tasting, high-quality wheats, which have resistance to the diseases and insects that are presently causing issues?"

One answer, he says, lies in localization, returning to the cultivation of crops suited for particular areas. North Carolina, for instance, has the largest acreage of wheat in the United States east of the Mississippi River. But, David says, "the types of wheats that are adapted to the South don't perform well in the Great Plains, and the Great Plains wheats do not perform well in the Mid-Atlantic states."

Wheat—like coffee, wine and truth—is multiple. In the United States, wheat is distinguished by season, color and hardness.[83] Hard wheats, such as bread grains and durum, have the most gluten and, therefore, the highest protein content. (Gluten is a set of proteins found in grains—including wheat—that impacts the elasticity of dough and gives breads and other gluten-rich products their structure and chewiness.) Hard wheats are dominant throughout America and largely grown in Kansas, Nebraska, Oklahoma and Colorado.[84] Soft wheats, which are primarily used for cakes, pastries and biscuits, are grown in the South.

Understanding that the more popular market in the United States is for hard wheat, local growers in Maryland, Virginia and North and South Carolina approached David and asked if there was a way they could expand into those types. "And my first reaction was ... no. But then I said we'll take a look at it. So we did." David and his colleagues grew samples of every single hard wheat available in the United States from the Florida panhandle all the way up to central Pennsylvania. But it was "off the shelf," he explains. "There was very little that had adaptation. And so we said, 'Let's see if we can do some breeding work and make selections.'"

Selections are neither easy nor quick: "It takes about anywhere from seven to 10 years when you first make a cross to when you have a variety that's available," David says. "You have to be persistent; the crosses we make today—those lines won't be ready for possible varieties until after

2020. And so we're always kind of thinking ahead as to what's needed. We made some crosses, we tested the material and some lines rose to the top." Two of the lines developed in 2009 did particularly well in North Carolina and Virginia and were released as premiere varieties of hard winter wheats by and for the East Coast.

"And they're very good wheats," David adds. "The quality produced on those is hard wheat quality that can be made directly into traditional and artisan breads—that's where Jennifer came in."

Jennifer is Jennifer Lapidus, founder of Carolina Ground, the only mill in the southeastern United States dedicated to the exclusive processing of regional grains. She was part of a small group of bakers who tested David's grain and deemed the flour was on par with grains shipped in from other parts of the country. "One of the long-term goals," David explains, "was to see if we could [cultivate viable grains], and I think that we have shown that. And then, in the grander scheme of things, to see if it greatly reduced East Coast millers' and bakers' reliance on trucking or shipping in through the railroad tons and tons of flour that's been produced in the central U.S." So far, that's working, too. As a result, David and other breeders have increased their number of experimental lines. "Those will be given to seed sellers," David explains, "to expand the options that farmers have." Choice was the cornerstone of this North Carolina wheat expansion—and Carolina Ground.

Jennifer and I met reluctantly. My car hinted at breaking down for the entire drive, and a journey that should have taken four hours took closer to six because (as usual) I got lost. The winding drive up the Blue Ridge Parkway toward the Appalachian Mountains that typically took my breath away and filled me with peace, instead, gave me vertigo. And, in a rush to travel to my interviews, I had forgotten to pack anything other than the old, loose sweatpants I had driven to Asheville in—and intended on sleeping in that night.

Jennifer, meanwhile, was at the end of a workday that had started at 4 A.M. Despite this, she kindly volunteered to pick me up from my motel at 6 P.M., but didn't arrive until closer to 7. The night was bitter cold and rainy; both Jennifer and I just wanted to crawl into our respective beds and call it a night. But instead, we headed to All Souls Pizza, a local restaurant founded by baker Dave Bauer, one of the people Jennifer had taken under her wing after he arrived in Asheville. We clicked; before the pizza made it to the table, we were swapping sips of wine and life stories like old friends.

Both Jennifer and Dave were students of Alan Scott, a blacksmith and co-author of the book *The Bread Builders: Hearth Loaves and Masonry Ovens*. His commitment to traditional baking methods helped spearhead a revival of brick ovens and pizzas and breads made on the fire—closest to the element that transformed them. "Alan's devotion was to three simple components," Jennifer told me, "the oven, the mill and the culture." The culture was *desem*, a natural starter added to dough to make it ferment and rise. It was a trinity of fire, earth and air.

Jennifer explained that she and Dave are part of a movement of breeders, farmers, millers and bakers trying to "understand the complexities in a simple loaf of bread." When I met Dave the following morning (still in my sweatpants), he said, "The whole model set up by people like Alan keeps you honest. You have to split wood and make a fire; it's tangible and real. And having to get up in the middle of the night is grounding. Staring through a gaze of flour dust is grounding."

Dave describes on the website for his bakery, Farm and Sparrow, that "in 2008, there was a wheat crisis. Prices soared, bakeries were closing, and the wheat berries I had been using disappeared. In their place, I began receiving pallets of wheat that were stale and hardly usable. No good. Around the same time, baker Thom Leonard sent out an email that he was going to release for sale grains and flours from a traditional American wheat variety called Turkey Red. I began using it, first bag by bag, then pallet by pallet. Eventually we said, to hell with it, this is our wheat."[85]

For Jennifer, "it was a higher purpose beyond the flour." The three components of her path became the farmer, the miller and the baker—"fostering the relationship between [them to] provide a tangible level of security and sustainability for all three."[86] It was also Jennifer's way of responding to the 2008 food crisis Dave referenced.

According to the World Bank, between 2007 and 2008 prices of food staples, including wheat, rose an average of 83 percent over the preceding three years.[87] This wasn't because of a global (or local) deficit of wheat but the result of the dynamics we learned about in coffee: Speculation on commodities markets distorted supply and demand and inflated prices to such an extent that many people couldn't afford wheat—or to eat at all.[88] Between March of 2007 and 2008, the average global price of the grain had climbed *130 percent*—a change also reflected in soy, corn and rice.[89]

The ecosystem of Carolina Ground, as well as other regional growers and bakers, reknits ties that globalization and industrialization have stretched thin—bringing the producer, consumer, and everyone in the middle, closer together. It offers us an opportunity for greater accountability, more intimacy and deeper connection. Jennifer continues to fight to make the mill "viable and sustainable," meaning a place that earns enough money to keep her from running herself into the ground and provides an alternative system outside of commodity wheat.

"I asked myself the question," Jennifer said over our final slices of pizza (made with a crust of local wheat and corn), "If every food tells a story, what story does our bread and our flour tell versus an industrialized loaf?" The answer is in the mill—and in the grain itself.

Flour isn't *just* flour.

The mill is where the grain becomes a good. What Jennifer is trying to do, by buying and processing local grain, is "to shift the production of grains from a feed-grade mentality to a food-grade one." That production becomes our daily bread.

"A field of wheat is converted into a loaf of bread by breaking the grain open and grinding it ... one of the common processes for making grains digestible and making their nutrients available to us," explains author Heather Wight. "However, mainstream flour production, for the most part, takes the nutritious grain and turns it into nutritionally poor flour."[90]

This is because of the way industrialized roller-milled processing separates the bran, germ and endosperm, the primary components of the wheat kernel or berry. The bran is the outer skin and contains the most fiber; the germ is the nutrient-dense embryo; and the endosperm, the inside portion of the kernel, makes up the majority of the wheat berry and is the source of starchy carbohydrates.

In order to remove the endosperm from the bran and germ, corrugated metal cylinders (roller mills) rotate inward and rub against one another at different speeds. The grain gets worked through roughly five sets of these roller mills to extract as much of the endosperm as possible. From there, the flour is sifted, bleached and enriched with a host of nutrients, some of which are present in the bran and germ but are set aside since the only portion of the grain needed to make flour is the starchy endosperm.

If the flour is labeled "whole wheat," this is the point when the bran that was initially removed is added back into the milled flour. The germ, which is rich in B vitamins, is thrown away. But then the flour is, bizarrely, refortified with some of the same nutrients the industrialized process took away, including fiber, potassium and calcium from the bran, and B vitamins, fats and proteins from the germ.[91]

After fortification, the flour is artificially aged. Aging improves the elasticity in bread and can be done naturally over a few months by exposing the freshly milled flour to air, or quickly by adding oxidizing agents like potassium bromate.[92] The final step is the addition of bleaching agents that turn the naturally yellowish-white endosperm a super-bright white.

The flour has now become fungible: all-purpose and interchangeable with any other bag of flour on the shelf. Its origin has become the grocery store.

This industrialization of grain and flour has happened slowly and insidiously. As brewmaster and author Garrett Oliver explains, "We knew back in the 1970s, as we walked through the supermarket, that something was wrong—but we didn't know exactly what it was. Somewhere in our hearts we knew bread was supposed to be made of just a few ingredients and probably go stale in a day. We knew it wasn't supposed to be full of ingredients we couldn't pronounce ... chemicals. We knew bread wasn't supposed to be Wonder Bread. People learned to take food and make it into food facsimiles: something that *reminds* you of food but isn't real food."

The alternative to this industrialized process is to keep the grain intact—to add nothing and take nothing away—in what's sold as "whole-grain flour." Carolina Ground grinds local, heirloom grains in small batches "with the germ crushed into the endosperm flour, spreading its oils and flavor." When sifting, Jennifer says, Carolina Ground "removes larger bran particles but doesn't fully separate the three components of the grain berry. What we offer to the bakery is a distinctly different flour, with flavors not detected in a roller-milled product." Chef Dan Barber clari-fies, "Using freshly ground whole grain flour allows you to taste things in the grain you otherwise would have missed. Each new grain—even each new *harvest*—we receive is very different. So, in that sense, there is very much a terroir to bread—a story that's being told not only about the soil but also about a larger community of breeders, millers and bakers."

While Dave slid loaf after loaf into his wood-fired oven, he reminded me, "As one of the world's first domesticated crops, bread is a symbol. What do we want to feed ourselves? I bake the way I do because this is the way I want to eat." And this is the way he feeds others, sating a hun-ger that goes beyond physical nourishment.

Originally from Wisconsin, Dave relocated to Asheville after nearly a decade in Minnesota. The traditional Southern foods he came across in North Carolina intrigued him. "Being a stranger in a strange land," he explains on his website, "I decided to use the bakery as a way of exploring these foods."[93]

I met him at his modest ranch home just outside of Asheville. I had expected a bakery in the middle of a residential neighborhood to stand out, but it, like Dave, was subtle. Aside from a few white towels hanging on the line outside, there was nothing that indicated his sweet blue home was a bakery. The portion of his house that had once been a garage now held a wood-fired oven and metal shelves full of warm loaves of bread. They were made from heirloom wheat and spelt that had been milled into flour just steps away from his house, in a building that was once the shed.

Dave's hands are in every loaf he bakes: seeded, spelt, heirloom wheat and, my favorite, heirloom grit. "Sometimes, in working with wheat," he said, "I feel more confused because I start seeing so much more. The bread becomes bigger and I become smaller." To that point, the bakery has now taken over the entire house.

A fellow traveler on Dave's exploration has been Tara Jensen, an artist and baker who assisted Dave and eventually took over the bakery he once occupied—one that Jennifer converted into a bakery in 1998.

I "met" Tara before I physically saw her, through photo after photo on Instagram (where her moniker is "@bakerhands"). I never realized pictures of bread could captivate me, but each image was intimate: the flour, water and salt; the bubbling, yeasty fermentation starter known as the mother, passed down from baker to baker and known as the source of bread flavor.

Tara and I spoke in Dave's living room, which is just off the bakery. We bonded over samples of his bread dipped in olive oil and our shared morning ritual of consumption of both coffee and horoscopes. She admitted she had been nervous to meet me because I had asked to interview her about the loss of agrobiodiversity in wheat. What she expected were questions on science, not star signs. I explained that, to me, genetic and cultural erosion are a bit of both. And then I asked her why she baked. "The reason I do this? I think it's important for us to have something to fold our life around. For some people, it's family or marriage. For me, it's bread."

With that, Tara handed me the plate full of bread. I reached for a small piece of heirloom grit: The crust was speckled with dry corn meal (grits), and the dough had cooked grits blended through. "This bread gives us a sense of place," she said. "It's an anchor, not just for one person but the whole community. It gives you a sense of where you are when you eat it. You're eating something someone made with his hands. In this one small thing, you're getting a lot."

I bit into the soft tip of the warm slice: It was like no other bread I have ever tasted before or since. It tasted like my childhood in North Carolina—toasted corn on the cob, morning grits, creamed corn—all there in this doughy mass edged with a crisp, hard crust. A taste that came through Dave's hands—and passed into mine.

Touch is the sensation of relationship. The weight and texture of a food—or the dish in which it's served—changes how we perceive it. Stale and fresh biscuits served in dishes coated with sandpaper are both considered crunchier than those served in smooth dishes;[94] yogurt served in heavier bowls is deemed richer and more expensive than the same yogurt served in lighter containers.[95] The intense burn of chilies on our tongues reduces our perception of sweetness (not sourness),[96] which, surprisingly, is also what happens when liquids become thicker in a smoothie or milkshake.[97]

Airy, fluffier bread tastes saltier than dense bread, regardless of the amount of sodium present: "By collecting samples of their subjects' spit, researchers determined that bread with larger pores releases its sodium faster when it's chewed. It's that rush of sodium that makes salt a mouthwatering saline sensation—even if there's less of it."[98] These experiences start with someone else's touch—touches that make the food or drink—but, most importantly, they are, ultimately, defined by our own.

After Tara and I parted, we exchanged a series of emails—notes of solidarity as we moved through jobs and breakups, and grappled with how to manifest what we cared about. "It's good that people are drawing attention to stone-ground versus roller-milled processing," she wrote, "and it's even better that testing is being done to discover the health impacts of pesticides rather than just focusing on gluten—but, really, what I want to see and participate in is a conversation on spiritual deprivation, lack of basic human self-determination and food security. You hit the nail on the head when you mentioned intimacy. I am afraid it is quickly disappearing along with all our heirloom varieties of grain."

Dan believes this is why "we're seeing such a backlash against grains and gluten." But, he explained to me, "we can't just cut them out of our diet. Instead, we need to rethink grains as a whole—how we grow them, how we grind them, how we consume them. That sounds unlikely—asking people to reimagine our ways of farming and eating—but I think it's where we're heading. Just look at what's happened to coffee in the last 15 years. Today, baristas and home brewers are not just religious about grinding their own beans—they also are conscientious about where their coffee came from, how it was grown, and when and where it was roasted. That's the future of baking. But the effects are going to be a lot more far-reaching."

These responses from Tara and Dan made me think of Dave in the shed, milling grain, and in the garage, stoking the fire and shaping the dough. Of Peter nurturing the grapes that Scott made into wine. And the Yeastie Boys—Chris and Steve—conserving the tiny pipettes of yeast that help produce beer and bread. I remembered Tadesse and Tebeje laughing as they showed me their worn hands. And the women at the Golden Temple *langar*, commanding the kitchen as my grandmother once had.

"The world touches us in so many ways," poet David Whyte reminds us. "We are something for the world to run up against and rub up against: through the trials of love, through pain, through happiness, through our simple everyday movement through the world."[99]

How do we preserve and honor this intimacy? How do we respond to the steady and growing drumbeats of climate change and crop failures? Of shrinking varieties of seeds that force us to rely on increased numbers of inputs? Of markets and monopolies that prioritize profits over people?

The short answer seems to come down to spending more, but it's really about reconnecting. Returning to one another. Not just for the earth or farmers we may never meet, but for *us*. When I reassured Tara that my exploration of agrobiodiversity was a blend of science and culture, I meant it. The loss of diversity we are experiencing—in wine, chocolate, coffee, beer, bread and beyond—is a reflection of something deeper, a loss of something essential to who we are. How we eat is part of the bigger mosaic of how we live.

"I love the practice of baking," Tara says, "but in the end what I do, what my work consists of, is other people and the relationships we have with each other. Bread is the vehicle through which these connections are forged and upheld."[100]

First, the relationships—and then the taste.

After my travels to Ethiopia, India and England—and before my final trip to Ecuador—I mailed Tara a dog-eared copy of a book on the relationship between food and faith. She, in turn, sent me a link to a piece she had written about her grandfather, the man who had gifted her with the scale she used to weigh out the ingredients for the first loaf of bread she ever baked. "It was collecting dust on top of his fridge, straining to feel the weight of something against its plastic body. Further down on that same fridge was this quote: 'Be gentle when you touch bread. Let it not lie uncared for and unwanted. So often bread is taken for granted. There is so much beauty in bread: beauty of sun and soil, beauty of patient toil. Winds and rains have caressed it and Christ has blessed it. Be gentle when you touch bread.' "[101]

After her grandfather died, Tara wrote that she took Communion, recalling that "the significance of bread to my grandfather came in the form of sacrament. Each week as a Eucharistic minister he helped to serve the body of his savior to his parish. 'For I received from the Lord what I also delivered to you, that the Lord Jesus on the night when he was betrayed took bread, and when he had given thanks, he broke it, and said, "This is my body which is for you. Do this in remembrance of me." ' It was bread in which he gave thanks and it was in bread that he tasted his salvation."[102]

That was it. That was the reason I had traveled to the Golden Temple. That was why even poisoned wheat held the potential of something sacred, and why the humble grain touched me like no other. We celebrate and mourn with it. We live through it.

Until I visited the Golden Temple, I thought *karah prasad* was holy because of the pandits who served it. Then, after my night with the *halwais*, I assumed it became holy through the prayers chanted over it. In both instances, I was wrong.

There are two ways to receive the edible blessing. The first is when passing through the inner sanctum of the Temple. Every person—regardless of their status or story—will be given the same type and amount of *prasad*. The second way to receive it is by lining up and making a donation. That portion is sliced with the *kirpan* and, over prayers, divided into two. Half of the thick pudding is given to the devotee, while the other half remains at the Temple to be disseminated among all who visit the shrine.

The moment *karah prasad* becomes holy is when it is shared.

When I first met Tara, I was slightly taken aback. She stood out against her landscape the way I, the *desi* girl in a Southern town, had in mine. But the quirky woman with the squared-off glasses and cute crocheted hat, with arms and legs full of tattoos and a history in anarchy, found her way and forged connection, as I, in my own way, had also done. She told me about eating Wonder Bread sandwiches with her work colleagues, the ones who thought the bread Tara baked was too dark and too sour. And that was okay, she said. "Because this is breaking bread, too. This is also communion."

My hope is that the winding, long journeys of our life staples will inspire changes in how you consider food and farming, as they did for me. But more than anything, I hope they serve as reminders that conservation manifests through cultivation, consumption and communion. Every encounter holds seeds of transformation. Every bite and every sip we take are our prayer.

III. TASTING BREAD

Love doesn't just sit there, like a stone, it has to be made,
like bread; remade all the time, made new.

—URSULA K. LE GUIN, *THE LATHE OF HEAVEN*

This tasting guide was developed by baker and author Michael Kalanty, who explains that he was drawn to baking because "it was the only part of the kitchen I hadn't worked in yet." Michael began developing his bread guide 15 years ago, when he was an instructor at the California Culinary Academy in San Francisco. The first version of the work served as what he called "a compass for students of bread." He created it, he said, "because the students were, like me, children of American bread culture. You know, soft pretzels from the mall and bread that came in white bags from the grocery store." Michael didn't develop a wheel. Instead, he created a graphic (found on page 350), which accompanies his guide.

I love Michael's protocol precisely because it is *not* the one I would have written. I am someone who loves coffee, chocolate, wine, some beer and all bread—but I am not a baker. I couldn't have told you, as Michael told me, that we start bread evaluation with the inside of the bread because "that's the heart of what the baker put into it." This is the journey of someone who has worked with dough for decades—and loves it in a way only a baker can.

One orientation isn't necessarily better than the other, but it reveals a focus one can only know by doing. Tara Jensen explains, for example, that "the crumb of your loaf tells the story of where your bread came from and of who you are." Every perspective—especially yours—matters.

You are ready to be your own expert. So find a quiet space, and grab a notepad and a palate cleanser (ideally, warm water). Bring all you are to this moment. You will find the touch of the maker and the farmer there, in every bite. As you take it into your body, and experience it through each of your senses, it now becomes part of you.

This bread is your bread.

Aroma and Flavor Notes for Bread, by Michael Kalanty[103]

First, you'll need a few slices of bread. Something from a bakery, with fuller flavor, will be more fun, but any type will do. If you are tasting two or three different kinds of bread, start with a lighter bread and then move toward the dark or dense ones.

Separate the crust from the crumb and set the crust aside. When tasting bread, you start at the inside, what baker's call the *mie* (pronounced "me"). From a technical perspective, the *mie* reveals the baker's skill in evoking the grain's natural flavors.

Chew your sample 10 or 12 times and spread the flavors over your palate. Chew with your mouth open, aerating the flavors as you chew. Much to the dismay of moms everywhere, the more air in your mouth, the more flavor you get from your food. Start by tasting for the sourness in the crumb. Then try to identify the character of that sourness. Does it have a smooth, dairy character like yogurt? Or does it have a sharp, tangy character like grapefruit? As you chew, look at the key at the bottom of the bread chart on page 350. For example, the sour/fruity family includes flavor notes such as green apple, grapefruit and lemon. All these produce a tangy sensation on the palate.

Complete your tasting of the crumb by focusing on its dairy sweetness. When baking lean doughs, such as baguettes (with low amounts of fat or sweeteners in the dough), it's not unusual to perceive an aroma of cow's milk or butter. A well-fermented baguette dough can smell, taste and sometimes even feel as if butter has been added to it. Depending on the flour and other ingredients in the dough, the aroma can sometimes seem like diacetyl, the butter-like flavor on some theater popcorn.

Now it's time to taste the crust. The crust is a different sensory experience altogether. It's dark, toasty and chewy. It's a study of contrasts when you sample the crust separately from the crumb. Caramelization of the sugar–amino acid compounds on the dough's surface brings notes of caramel, nut and brown butter. (Many of those notes you associate with the Maillard reaction; the bitter/sweet richness of a grilled steak is an example.) Oven temperatures aside, two things contribute to the flavors of the crust. One is the amount of whole grain flour in the bread. The other is the total fermentation time of the dough.

The crust of a white flour bread, like the baguette, has the simple character of toasted grain. Sometimes there's a warm dairy note, too, like the one associated with condensed milk or toffee. When a dough contains whole grain flour, there's a malted character to the crust. Take the classic French sourdough, *pain au levain:* The crust reveals notes of stout or dark beer with a sour undertone. The other factor contributing to crust flavor is the total fermentation time of the dough. Short-fermented, same-day breads have a toasted-nut character, like almonds. Long fermentation encourages sugars to accumulate in the dough. As these caramelize, they display the meaty character of grilled steak or roasted mushrooms.

Words like "chocolate" and "balsamic" aren't traditional descriptors for crusty bread, but, to our surprise, many heritage grains are delivering notes like these, and more. As our grain economy becomes more localized and these flours find their way into our kitchens, there's opportunity for greater sensory experience.

Within minutes of blending yeast, water and flour, fermentation begins. It continues relentlessly until the dough is baked and its internal temperature reaches 122 degrees Fahrenheit or so (the thermal death point for yeast). As long as the dough is active, it continues to ferment. It is both the art and the science of the baker to guide this fermentation. Done well, the finished bread yields aromas of whole grains—earthy and nutty with a soft mineral note, like flint. Uncontrolled, the results can be pithy grapefruit bitterness, "turned" red wine or sherry.

Assess the grain character. A separate component of bread tasting includes a discussion of the fermentation character of the grain itself. There are four phases of grain fermentation, each corresponding to different sets of flavor and aroma compounds in bread. These subcategories group the more common aromas and flavors found in breads.

- *Simple.* Non-yeasted doughs need no fermentation. Indian chapati and Mexican tortilla doughs are mixed and receive only a brief rest before shaping and baking, mainly to relax their elasticity. Some yeasted flatbreads like naan and pita receive a short (by Western standards) fermentation. The toasty, nutty and sometimes charred notes of their flavor profile come from the cooking method. The dough itself can display notes of straw, raw starch or a dry yeast aroma.

- *Moderate.* As yeast enzymes break down carbohydrates in the flour, simple sugars are released into the dough. As it eats these sugars, the yeast burps out carbon dioxide, in turn, making the dough rise. During this stage, the raw, starchy flavors of the grain decrease. Aroma and flavor notes associated with other "cooked" or processed starches take their place. Midway through the first rise, the dough displays notes of al dente pasta or steamed potatoes. A yeasty character appears, like that found in certain Champagnes. A good reference for this stage is the plastic-wrapped white sandwich bread you find in grocery stores. This is considered a moderate grain character.

- *Complex.* At a certain point, the yeast activity levels off. More of the carbohydrates have been broken down and the aromas and flavors of the digested starch evolve. The dough delivers the earthy character found in cooked whole-grain cereals or dried beans that have been presoaked and cooked. If a baker used a cool fermentation process and refrigerated the dough overnight, the yeasty flavor is replaced with notes of green olive and a flint-like mineral note. Bakers consider this a complex grain character. The culinary reference would be a hearth-baked ciabatta.

- *Over-fermented.* Left unchecked, it's possible for the grain to over-ferment. Even under refrigeration, a lean dough shows aromas of over-fermentation somewhere around 72 hours; more time isn't always better. The aroma and flavor notes displayed in an over-fermented dough include beer, a grapefruit bitterness, and the aroma of red wine that has "turned" and is becoming vinegar.

"Every technique a baker uses delivers its own character to the dough," Michael explained to me. "Taken together, they leave an immediate and identifiable mark, a kind of fingerprint. Every bread has its own style—its own soul." As he said this, I thought about every sense we have explored and the ways in which this guide and the array of wheels have led us in our culinary discovery. The mark of the baker, the farmer, the miller—the soul of what they become—that is also a part of flavor.

Now do this with everything: *Feast on your life.*[104]

OCTOPUS

I. EIGHT TENTACLES

I scanned the Peruvian customs form, similar to ones I had completed an endless number of times, but paused when I got to the question "Have you been on any farm?" The boots on my feet that had carried me from wild coffee forests to cacao plantations had been scrubbed clean. Bags that once carried heirloom seeds were, instead, stuffed with notes from my interviews on biodiversity. I checked the box "no," despite knowing full well I carried each and every one of those places in and on me. Wendell Berry echoed through my head: "Eating is an agricultural act."

Research shows we make about 200 more food decisions each day than we are aware of, most of which we consider "automatic," informed by far more than what we are actually hungry for, including the time of day, the environment in which we eat our foods, and the company we keep.[1] Associate professor of psychology Katherine Appleton summarizes, "Few of us eat just because we are hungry."[2]

This recognition was solidified for me at the grocery store, the place where we started our journey and the first destination I seek out when visiting a new country. The areas where food is sold reveal a lot about a culture and what it values. "Tell me what you eat," proclaimed French epicure Jean Anthelme Brillat-Savarin, "and I will tell you who you are."

Who I was had changed: I wandered through the aisles at the Vivanda grocery store in the Miraflores neighborhood of Lima, Peru, eager to try new things. It was winter of 2014, and I was in Peru for a forum on landscapes, development and climate change, excited to discuss agricultural impacts. Instead of toting my usual culinary touchstones (Ethiopian coffee, Aeropress coffeemaker and multiple bars of chocolate), all I wanted to do was try the new.

How did this happen? How did I move from safety to adventure? Which people, places and foods prompted the shift? I'm not sure. All I know is that it happened one meal at a time. In the beginning of this book, I mentioned a study that asserts the way we get over our fear of new things is by trying them. But when I first typed those words, I wasn't there yet. I was still living into that answer.

Although the romantic relationship I had hoped for (the one that prompted me to smoke and left me sad after sensory training) didn't pan out, a different kind of connection had taken hold—one that allowed me to reach toward others and ask, "What are you eating?" This curiosity resulted in one shopkeeper closing her store early to walk me to the bakery she loved because she couldn't bear the thought of me having a subpar baguette; and another to cook for me because she knew it brought me joy. It's the connection that comes from knowing we are more alike than different—that to share a meal or a table is, in some small and meaningful way, a means to setting differences aside and joining together.

This journey was harder than expected. I travel—a lot—but my culinary adventures had always been constrained. The first time I tried to have a nice meal by myself was when I was 16. I joined my mother at a work conference in Myrtle Beach, South Carolina. She had an evening function, so she handed me her credit card and steered me toward the nicest restaurant in the hotel. I had on my fanciest dress and my most mature affect, but the staff wasn't buying it. They seated me—the kid—in the corner by the bathroom and ignored me for most of the evening.

More recently, in Orvieto, Italy (as I started my research for this book), I arrived at a restaurant at 6 P.M.—long before any self-respecting Italian would ever show up—and shakily announced, "Per una, per piacere."

"Per una?" the handsome waiter asked incredulously. "Perché? Sei molto bella!" I smiled weakly. Beauty had nothing to do with it. And, yes, I wished it was "Per due, per favore," but it wasn't. Italy is for lovers: *per due*. The waiter was overly attentive, which was only slightly better than being ignored: His doting highlighted my solitary state for the handful of tourists also eating at that early hour. While the *bucatini all'amatriciana* was prepared to perfection, it was overshadowed by my self-consciousness. I never made it to dessert.

Fast-forward three years, when I, somewhat reluctantly, booked a table for one at a highly recommended *cevichería* in Lima. It was my only meal away from the demands of the presentation to which I'd committed, and I wanted it to be good. Nothing had gone as planned with the work gig: I had spent 28 hours in transit in order to spend 51 hours in Lima, and the speech I had labored over for two weeks was edited down to one minute. I needed a break—and I wasn't going to settle for room service. *Today, I will drown my sorrows in a pisco sour,* I wrote in my diary.

I arrived with high hopes and an empty stomach to the restaurant El Mercado in the Venice Beach–like neighborhood off Avenue Hipólito Unanue. It was beautiful in a rustic/industrial/hipster sort of way. I walked through the wooden entryway, past the long line of people without reservations, to the small table where an affable brunette hostess stood. She smiled, checked off my reservation—and ushered me to the worst table in the restaurant, right next to the drink station with a view of the sink. I withered and shook my head. "No."

"¿Cuál es el problema?" the hostess asked.

"No. Imposible," I said. I wasn't going to settle.

She raised her hands and swept them around the restaurant. "Es lo que tenemos." *It's what we have.* I followed her gesture with my own, jabbing in the air at several open tables. She tried to explain they were for bigger parties and suggested a seat at the bar, explaining I could eat the full meal there.

"No." I was holding firm. I could clearly see there were nicer two-tops than what she had offered. My table for one would have to be better. She gave me a "Well, you can leave" kind of shrug and turned her head back toward the line of people queued in front of the hostess stand. I felt a little betrayed by the absence of sisterhood and felt myself starting to judge her acid-washed jeans. But then I stopped myself, smiled and said, in horrifically mangled Spanish, "I am a journalist writing about food. I made a reservation weeks ago. I called from the United States. I want to be here."

She sighed—and relented. "Sígueme." *Follow me.* She led me to the beautiful, upholstered area in the middle of the restaurant that had *three* vacant two-tops.

This small victory was a defining moment for me because part of this journey has been about remembering who I am and what I deserve. I deserve a nice table for one. We all do. It seems obvious, but it's a lesson I'm still learning.

People are often incredulous that someone who has traveled all over the world sits in her room and orders room service.

I do. I *did.*

Now I go out.

The waiter was patient and kind as I fumbled through my order. I asked, in Spanglish, for half portions of three of the best dishes on the menu—*ceviche, tiradito* (raw fish in a spicy sauce) and *pulpo* (grilled octopus)—plus the signature drink of Peru, pisco sour. He assured me I had chosen well and would enjoy it all.

"Sí, verdad," I replied. "Everything is new to me. I *will* enjoy it all."

No sooner had those words come out of my mouth than I realized I had changed. I was no longer apprehensive. I was now actively moving toward the unfamiliar—not out of a sense of sensorial or journalistic duty—but because I was excited to.

Twenty minutes later, the waiter returned with three mouthwatering dishes, but one, in particular, captivated me: A small eight-tentacled, deep purple wonder, the octopus was gorgeous. Cooked over fire, it was served with grilled mushrooms, small tomatoes and capers. A few days prior, I had eaten what, until that point, was the most delicious octopus of my life at a wine bar in Charlotte, North Carolina. That octopus had been slightly rubbery, softened by a quick steam in a pressure cooker. This one was the opposite, with flesh so tender I could cut it with my fork. The dish was slow and deliberate. It tasted of wood and the ocean—the very essence of *pulpo*. It made the defeats of the day and discomfort of dining alone worth it.

It made me cry.

Italian scientists from the University of Naples have discovered that when we eat for pleasure, rather than solely for hunger, the reward circuit in our brain—that primal neural network that fires when we win a prize or have an orgasm—is activated. This reinforces our desire to reach for foods based on how they taste versus strict caloric nourishment.[3] It is a desire that, literally, transforms how we receive food. In one study, clinical nutritionist Leif Hallberg and his colleagues at the University of Gothenburg in Sweden found that people absorbed significantly more iron from food (when roughly the same amount of iron was available), depending on the composition of the meals.[4]

This was evident in the *pulpo*. It had been made carefully and was a completely different experience from other octopuses I had eaten. Like my cup of Yirgacheffe coffee, the octopus invited me to pay attention and savor. It nourished me deeply.

What is essential isn't limited to nutrients. What is essential is what gives us pleasure—and from where we derive meaning. Food transcends the mundane and interweaves the disparate. It reminds us of our interdependence. We are what we eat, and we eat what we are.

With each bite of octopus, tears welled up in my eyes. I couldn't stop them, but, trust me, I wanted to. I definitely didn't want to be the strange single lady in the middle of the restaurant crying over a plate of food.

My tears were mostly about the exquisite taste, but some small part of those tears was made of the sad acknowledgment that, for years, I had been depriving myself of this. In city upon city, I had chosen to sequester myself away, dining on room service and takeout, waiting until one became two.

Not anymore. *No más.*

When had I become someone who cared so much? This person who, through daily cups of coffee and occasional plates of octopus, started to see both the world and herself in their composition? When did my reflection begin to show up in the glass of wine? As my pisco sour buzz set in and my tears waned, I started to realize this evolution was always happening. Only now, I was paying attention. Our tastes, our food preferences, our environments, the flora in our guts—they are always changing. Change is the constant, and when we become conscious of it, something amazing can happen. We can shape it and love it; we can make it our own. We, the eaters.

I pulled out my journal and started writing: *It's Sunday. I am in Lima, and I am not alone. There is jazz on the stereo. The old ladies are laughing, the young kids are playing and there, in the corner, is a married couple still in love. I want to open up to all of it. I want to go from table to table and ask, "Have you tried the pulpo?" I want to tell them it made me cry. I am slightly drunk; I think the whole restaurant is slightly drunk: on pisco, on life, on love. I feel sated. I feel joy. And, for that, I am grateful. I am so obviously alone–and I am so obviously not.*

As I scribbled away, the chic woman at the table next to me, whose ankle boots I had been admiring for nearly two hours, leaned over and asked, "Are you a food critic?" No, I said. But I write about food.

This never happens to me. I never get chatted up. And then it happened, again, on the way out. As I left, I ran into people who had seen the one-minute speech I traveled 28 hours to give. They commiserated with me. One offered me a cigarette. I accepted. All was well.

To most, a solo meal isn't a courageous act. But to me, it was, because it revealed my vulnerability around being alone and, in being by myself, feeling like I was settling. Now I know I'm not. I was not actually alone: I was *with* myself, having one of the very best meals I have ever tasted, surrounded by people celebrating the same. That meal opened up the world by reminding me that—multiple times a day, with every food or drink that passes my lips—I am connecting with others. My meal would not exist without the people who caught the octopus, the farmers who raised the vegetables, the chef who tended to the grill and the waitperson who brought it to my table. And nothing would exist without the bigger web of life that sustains the diversity we are lucky enough to eat. This holds true for every meal. We are never alone.

I forget this every day, only remembering when I slow down and show a little courage and do what's uncomfortable—or show a little gratitude and feel that grace. In the end, that's all we can do. Let it all in. Savor everything. Scrape the pot, lick the spoon, dig our hands into the messy fruit. Twirl it around on our fork, swallow every drop, find the sweet in the bitter. Smell, taste, touch, hear, feel—and ask for the better seat.

This is not to say I am walking around with some majestic sense of unity every time I eat a meal. But it is to say I now know every meal, both humble and exotic, connects me to a greater whole. And that these stories of our foods and drinks can change; we can create new narratives. This is why I choose biodiversity: foods that are slow, not fast; different, not the same. They are the manifestation of deeper human connection. They nourish and nurture. They are foods of love—and they reflect who and how I want to be.

"Live deep," Henry David Thoreau once wrote, "and suck out all the marrow of life."[5] It's there. And what it will ultimately do is bring you joy. Or maybe—in a bustling café over a plate of seafood—transcendence.

This is a book about food, but it's really a book about life.

II. THREE HEARTS

I wanted the book to end there, poetically, with encouragement to live deep and create the narrative you wish to see. But this is a story about life, which means there is always a new chapter and another lesson.

Last spring, Dutch chocolate seller Erik Sauër said to me, "By sharing stories, we become a part of them." The story of agricultural biodiversity now belongs to you—to shape and manifest through your own hands and palate. You can transform the world by trying different varieties of chocolate, drinking a diverse range of beers, exploring the terroir of bread—and extending that same culinary curiosity into your life staples. But there's one place where it makes greater sense to do the opposite: the ocean, something I didn't know when I was crying over my octopus in Peru.

I, too, am learning.

In trying to eat virtuously, I had opted for fish and other marine life over animals, like cows (which use 11 times the water of pigs or chickens and 28 times more land, including the amount of farmland used to grow cattle feed).[6] But that decision failed to consider the waste generated through fishing. The trawlers used to harvest the catch of the day are the equivalent of bulldozers that clear-cut forests: They take out everything.

"Imagine being served a plate of sushi," Jonathan Safran Foer writes in *Eating Animals*. "But this plate also holds all of the animals that were killed for your serving of sushi. The plate might have to be five feet across."[7]

The biodiversity contained in our oceans is like that contained in our soils: vast and limitless. Fish and other marine life are our last wild foods. "Our origins are there," explains oceanographer Sylvia Earle, "reflected in the briny solution coursing through our veins and in the underlying chemistry that links us to all other life."[8] And we are only aware of a fraction of life contained within those waters: "Nearly every inch of Earth's lands have been mapped. ... Meanwhile only about 5 percent of the ocean below the surface waters has ever been seen at all, let alone fully explored."[9]

I met Sylvia seven years ago when I interviewed her for a television program I used to host. I had just finished her book *Sea Change: A Message of the Oceans,* and the first question I posed to her was brimming with anguish: "Dr. Earle," I implored, practically wringing my hands, "over 90 percent of our fish stocks are depleted. What do we do?" She smiled kindly and said calmly, with full compassion, "The other 10 percent, Simran. We fight to save the other 10 percent."

Every single time I have fallen into a well of despair, that response from Sylvia has pulled me back out. We fight for the other 10 percent, for the biodiversity we *can* save.

Our oceans are the watery equivalent of forests. They contain 50 to 80 percent of all life on Earth and contribute to every breath we take.[10] (Microscopic phytoplankton that live near the surface of water release 50 percent of the world's oxygen when they convert sunlight to energy.[11]) Oceans are the lungs of the world. It would make sense to preserve the places that, literally, enable us to live, but only 1 percent of oceans and adjacent seas are protected, compared to 12 percent of land surfaces.[12]

The octopus I devoured belongs to the stocks of marine life that have been overharvested and are not being conserved or replaced. The best way to save biodiversity in the ocean, Sylvia explains, is to stop eating these exotic animals and to stick to the mundane: the smaller fish that readily reproduce. If we don't, studies project the "global collapse" of all saltwater fish by 2048.[13]

All. Extinct. Forever.

I learned this during a series of interviews I did with Sylvia in Australia. Sylvia was the first woman to walk on the ocean floor and has logged, in total, over 6,500 hours underwater. In the same year I was born, Sylvia led the first all-female team of aquanauts on expedition. She has dedicated her life to the ocean—and has not lost hope. So, despite the punishing deadline for this book, I jumped at the chance to interview her once again.

In our first talk, I described how I had grown to love octopuses, without being specific that what I loved was *eating* them. Her eyes lit up and she described them as "curious and delightful." As she spoke, I started to feel a different love emerge. Blue blood courses through octopuses' three hearts. Two octopuses mate to create about 56,000 eggs, but roughly two make it through pregnancy.[14] Ed Yong writes, "For many a female octopus, laying eggs marks the beginning of the end. She needs to cover them and defend them against would-be predators. She needs to gently waft currents over them so they get a constant supply of fresh, oxygenated water. And she does this continuously, never leaving and never eating. When the eggs hatch, she dies, starving and exhausted."[15]

The oldest octopus fossil in existence belongs to an animal that lived almost 300 million years ago.[16] "They have been around for such a long time," Sylvia said with a combination of awe and sadness.

It was time to fess up. I admitted my love was octopus on the plate, not underwater—to which Sylvia gave another kind, calm and fully compassionate response: "If you have to eat marine life—or meat, for that matter—eat the ones that eat plants. Don't eat the fish that eat other fish. Eat tilapia—lots of it—but not the tuna, not the lobster, not the carnivores or ancient creatures. Think about all the pollutants [we've dumped into the oceans] that they've absorbed—and all the animals they've eaten in the course of their long lives. And remember, once we lose them, they're gone."

I have spoken about heartbreak and loss throughout this book. It has defined the last several years of my life in ways that have been devastating, for sure, but also empowering and unexpected. Empowering, because what doesn't kill you really *does* make you stronger—and softer and more compassionate about the suffering of others. Unexpected, because it hasn't just been limited to me and another person. My heart has broken for the ways in which we, as a society, have slowly but surely eroded who we are, as well as what we need to sustain life on the planet.

"Realizing its inescapable nature," poet David Whyte writes, "we can see heartbreak not as the end of the road or the cessation of hope but as the close embrace of the essence of what we have wanted or are about to lose."[17]

"Yet we use the word heartbreak," he continues, "as if it only occurs when things have gone wrong: an unrequited love, a shattered dream. ... But heartbreak may be the very essence of being human, of being on the journey from here to there, and of coming to care deeply for what we find along the way."[18]

My heart has not only broken, it's broken open.

This time, when I asked Sylvia "What do we do?" she responded with a question. "If this isn't what we want, what do we choose?" she asked. "We still have the ability to transform the system and heal ourselves. We have proven we're really good at killing things. Can we also demonstrate our ability to care, nurture and restore?"

The loss of biodiversity in the ocean (and on land) is directly connected to the decisions we make—from overfishing and pollution, to habitat loss and climate change. We have the choice and power to save foods by eating them—and, in some cases, by choosing *not* to. To find hope and to save the remaining 10 percent.

"To be hopeful," writes author Rebecca Solnit, "means to be uncertain about the future, to be tender toward possibilities, to be dedicated to change all the way down to the bottom of your heart."[19]

This is a book about life—and all the hope our hearts can hold.

METHODS OF CONSERVING AGRICULTURAL BIODIVERSITY

Agricultural biodiversity (agrobiodiversity) refers to the variety of life on Earth that is directly related to food and agriculture. It's what emerges out of the connection between

1. the microorganisms, plants and animals we eat and drink;
2. the inputs that support their creation and development, including bees and other pollinators, as well as the quality of nutrients in the soil;
3. nonliving (or abiotic) influences on our ability to grow and gather food, such as temperature and the structures of farms; and
4. a range of socioeconomic and cultural issues that inform what and how we eat.[1]

Although there are multiple reasons for the loss and gain of agrobiodiversity (as outlined in Appendix II), scientists, farmers, breeders, activists—eaters and drinkers from all over the world—are working toward conserving these diverse natural resources.

Biodiversity is preserved *ex situ* in stored collections, *in situ* in the wild and *in situ* on farms. These methods of conservation are detailed for easy reference:

Ex situ conservation: Latin for "out of place," *ex situ* conservation is the collection of plants and animals saved for their global or national significance. These genebanks, or germplasm repositories, include seed banks and collections of plant and animal genetic materials, such as sperm, embryos and microorganisms.

Examples of *ex situ* collections are the National Collection of Yeast Cultures (NCYC) in England, the USDA National Clonal Germplasm Repository in the United States and the International Cocoa Genebank in Trinidad.

In situ conservation: Latin for "in place," the two main methods of *in situ* conservation are to save and protect what's already growing in the wild ("*in situ* conservation in the wild") and to sustain a plant or an animal by actively growing or raising it ("*in situ* conservation on-farm").

An example of *in situ* in the wild is the Kafa Biosphere Reserve in Ethiopia.

An example of *in situ* on-farm is Alberto Bautista's cacao farm in Ecuador.

The final form of conservation discussed is *in vivo* ("in living") conservation—saving diverse foods and drinks by consuming them. It is what happens in our living and doing, by consuming in a way that reflects and contains our care for diversity.

DRIVERS OF CHANGE IN AGRICULTURAL BIODIVERSITY

Developed with Paul Boettcher (animal production officer, FAO), Luigi Guarino (senior scientist, Global Crop Diversity Trust) and Stefano Padulosi (senior scientist, Bioversity International)

The natural world is interdependent; everything within our ecosystem works in response to, or in tandem with, something else. Changes in agricultural biodiversity are almost always caused by the interaction between multiple drivers that are both direct and indirect, and occur simultaneously or intermittently. So while the drivers are listed here individually, in reality, they are dynamic and in relationship; one or more drivers of change impact the development, rate and intensity of others.

Direct drivers are the ones that cause shifts in biodiversity, while indirect drivers influence the rate or magnitude of direct drivers. There is overlap between the two. Global warming, for example, is a direct driver but could be considered an indirect driver that leads to changes in land use and habitats. It could speed up the development of diseases and invasive species, but might also create the opportunity for different breeds and varieties to thrive.

Direct drivers include changes in habitats or the way we use land, modifications in our diets, global warming, pollution, diseases and invasive species. Cultural, demographic, social and economic factors—plus political, scientific and technological changes—are indirect drivers. These drivers have the potential to not only reduce agricultural biodiversity, but also, in some instances, *increase* it.

Each driver can play a varying role and has a different impact depending on its context. Some drivers are in our control; others aren't. Some take generations to appear; others manifest over a short period of time. We don't yet know what our future holds, but we do know agrobiodiversity requires greater attention.

1. SOCIOECONOMIC CHANGES

- Increased physical and psychological distance between producers and consumers of food, greater disconnection from agriculture
- Changes in taste, demand and diet (including expansion of the global standard diet, increases in processed food consumption, declines in the diversity of food varieties, marginalization of foods that require elaborate preparation, and religious restrictions on certain foods)
- Increased wealth (which may result in a desire to consume more meat and adopt a Westernized diet)
- Changes in food availability and price
- Changes in culinary preparation (increased purchase of processed/prepared foods, reduced knowledge about cooking)
- Reduced time for food preparation
- Reduced generational knowledge of, and appreciation for, agrobiodiversity
- Decreased dependence on local foods or local inputs for agriculture
- Poverty, social inequity
- Population growth
- Urbanization, migration
- Changing livelihoods
- Increased female participation in work outside the home (women do most of the cooking, which may decline when they work in external settings)

- Industrialization
- Demands by manufacturers for more uniform raw materials for food (both decreased number of—and decreased variability within—species, breeds and varieties)
- Demands for higher production breeds and varieties
- Globalization
- Trade agreements
- Food and agricultural policies (such as regulations on seed trade and livestock breeding)
- Intellectual property rights/patents
- Plant variety protection systems and policies
- National sovereignty over plant and animal genetic resources
- Shifts in research and development in food and agriculture from the public sector to the private sector
- Policies and developmental assistance that prioritize high-yielding varieties (HYVs)

2. AGRICULTURAL SHIFTS

- Consolidation of food and agricultural supply chains
- Reduced and/or changing political clout and lobbying power of the agricultural sector
- Decrease in extension programs and other educational support services for farmers
- Replacement of heirloom and landrace crops with HYV seeds and breeds (such as hybrids, crossbred varieties) and transgenic (genetically engineered) crops that are not adapted to local habitats and/or are infertile, do not render true, and/or are patented, therefore increasing dependence on commercialized plant and animal resources
- Preference for high-output and/or single-productivity traits in varieties and breeds, which replaces diverse landraces in breeding programs and cultivation
- "Path dependence"—a construct in which decisions on future crops are determined by a small pool of currently commercialized

crops (e.g., food and agricultural choices in developing countries are determined by choices in more developed ones)[1]
- Emphasis on improving only one or a few breeds, varieties or species through genetic technologies
- Breeding programs and methods that prioritize "genetic response," changes in plant or animal populations from one generation to the next, over biodiversity
- Emphasis on standard agricultural practices that breed and/or harvest plants and animals uniform in size and shape
- Mechanization, resulting in or contributing to:
  - Erosion of indigenous knowledge and agricultural traditions
  - Increased size of farms
  - Homogenization of breeds and crops (monocropping of uniform breeds and varieties, reduction of crop rotation)
  - Degradation of soil
  - Reduced use of animal species and breeds for transportation
  - Increased application or overuse of inputs (including water for irrigation and petrochemicals for fertilizers, herbicides and pesticides)
- Expansion of biofuels

3. HABITAT LOSS AND LAND-USE CHANGES

- Deforestation
- Overgrazing
- Conversion of ecosystems
- Restrictions on the use of traditional grazing and cropping areas
- Large-scale land acquisitions (known as "land grabbing")
- Mining
- Logging
- Other destruction of natural habitats (due to floods, tsunamis, fires)

4. OVEREXPLOITATION

- Overharvesting from the wild (e.g., medicinal plants, fish)
- Coral reef collapse
- Development

5. CLIMATE CHANGE

- Loss or expansion of the "environmental envelope"—the collection of all environmental variables (including climate, soil and other organisms) that define or limit the places where a given plant or animal can grow
- Emerging diseases and pests
- Sea level rise (resulting in loss of habitats)
- Increased natural disasters

6. INVASIVE SPECIES, BREEDS, VARIETIES*

- Pests
- Weeds
- Microorganisms
- Introduction of exotic species that compete with, prey on, or mate/hybridize with native species

   Naturally occurring species, breeds and varieties can also be a problem

7. DISEASE

8. VARIABILITY OF NATURAL RESOURCES

9. POLLUTION

- Persistent organic pollutants
- Pharmaceuticals
- Heavy metals
- Fertilizers, pesticides and herbicides
- Plastic
- Acid rain, fog or mist

10. POLITICAL INSTABILITY

- Prolonged social conflicts and wars that can cause a complete loss of agrobiodiversity in the wild, on farms and in *ex situ* collections
- Amendments to and/or abandonment of genetic resource policies as a result of regime change

11. IMPROPER MANAGEMENT OF *EX SITU* GERMPLASM COLLECTIONS

- Diminished and/or inconsistent funding
- Mislabeling
- Poor characterization (or misidentification)
- Inadequate storage
- Interruptions in supply of electricity or liquid nitrogen (for cold or cryogenically preserved resources)

12. LACK OF AWARENESS ABOUT THE IMPORTANCE OF AGROBIODIVERSITY

INTERVIEWS

My deepest gratitude to all the people below, who not only shared their time and insights, but are actively engaged in making the world more diverse and delicious. I am a better person because of all that you shared.

INTRODUCTION / WHAT'S AT STAKE

Alice Julier, Barbara Herren, Ben Shewry, Braulio Ferreira de Souza Dias, Carolin Bothe-Tews, Cary Fowler, Caterina Batello, Colin Khoury, Devra Jarvis, Emile Frison, Erin Betley, Evelyn Kim, Gyorgy Scrinis, Jerome Waag, Leah Gauthier, Luigi Guarino, Mauricio Bellon, Michael Halewood, Michael Hermann, Pablo Eyzaguirre, Patricia Merrikin, Paul Boettcher, Paul Bordoni, Phrang Roy, Russ Jones, Stefano Padulosi, Stephan Krall and Terry Sunderland.

WINE

Ann Noble, Bernard Prins, Caleb Taft, Colleen McGarry, Deborah Golino, Nancy Sweet, Pax Mahle, Peter Fanucchi, Scott Schultz, Snow Barlow and Tala Drzewiecki.

CHOCOLATE

Alberto Bautista, Aldo Reyes, Alexandra Arces, Alyssa McDonald, Andy McShea, Anna Davies, Anna Laven, Antoinette Sankar, Brad Kintzer, Brigitte Laliberté, Carla Martin, Chloé Doutre-Roussel,

Colin Gasko, Cristian Melo, Darin Sukha, Daysi Rodriguez, Debra Music, Diego Badaró, Ed Seguine, Erik Sauër, Fanny Julieta Hurtado, Frances Bekele, Francisco Bienvenido Peñarrieta, Francisco Chiribog, Francisco Valdez, François Ruf, Freddy Amores, Gino Dalla Gasperina, Gisela Caicedo Bustamante, Gualberto Valdez, Gustavo Garcia, Javier Enrique Valencia Castro, Jean-Marc Anga, Jeffrey Stern, John Kehoe, José Valdivieso, Joshua Loomes, Jude Solomon, Karla Panezo Reyes, Katie Gilmer, Laila Mina, Lambert Motilal, Lee McCoy, Lourdes Paez, Luiz Javier Meza Cabrerra, Maria De Lourdes Alvear Loda, Marissa Moses, Mark Christian, Martin Christy, Maryuxi Espinoza, Matthew Escalante, Michel Boccara, Mina Bustamante de Caicedo, Mireya Vite, Naailah Ali, Nadia Rosales, Nathan Palmer-Royston, Pam Williams, Pathmanathan Umaharan, Pedro Romo-Leroux, Philipp Kauffmann, Red Thalhammer, Rena Kalloo, Rich Tango-Lowy, Robbie Stout, Samuel von Rutte, Seneca Klassen, Sergio Cedeño, Shawn Askinosie, Steve DeVries, Sunita de Tourreil, Surendra Surujdeo-Maharaj, Tad Lombardo and Vicente Norero.

COFFEE

Aaron Davis, Aaron Wood, Abbie Yunita, Adam Overton, Alazar Mengstu, Alemayehu Teressa Negawo, Alemu Seda, Andenet Bekele, Atmasu Lakew, Babur Damte, Balkew Tadesse, Bereket Beyene, Birhanu Zegeye, Caitlin McCarthy-Garcia, Christa Lachenmayr, Demelash Teferi, Eileen Kenny, Fantahun Beyene, Feyera Senbeta Wakjira, Frehiwot Getahun, Gemedo Dalle Tussie, Hagos Gidey, John Di Ruocco, Kayla Nolan, Kc Reynolds, Kurumi Ikaki, Mammo Mali, Markos Lole, Matt Ledingham, Mauricio Galindo, Melaku Addisu, Mesfin Tekle, Metasebia Yoseph, Michael Sheridan, Mick Wheeler, Mike Mamo, Mintewabe Monna, Moata Raya, Negussie Mekonnen, Nicolas Lawson, Peter Giuliano, Phyllis Johnson, Sarada Krishnan, Sisay Tesfaye, Stephen Vick, Tadesse Gudina, Taye Kufa, Tebeje Neguse, Tefera Roba, Telehun Atara, Tiglu Melese, Tim Schilling, Tsion Taye, Warren Langley, Werko Robe, Willem Boot, Workineh Mulat and Yohannes Gebre.

BEER

Adam Elliston, Ben Ott, Carmen Nueno-Palop, Chris Bond, Garrett Oliver, Javier Carvajal Barriga, Jeanette Newman, Kelsi Moffitt, Mark Dredge, Mike Ambrose, Paige Latham, Richard Boughton, Steve James and Todd Bates.

BREAD

Ajit Singh, Alok Gupta, Beti Minkin, Bhupinder Singh, Dalvar Singh, Dan Barber, David Bauer, David Marshall, Derek Mitchell, Devinder Sharma, Eli Rogosa, Guirvail Singh, Gurinder Singh, Gurlal Singh, Gyani Kewal Singh, Harpreet Singh, Jagdish Singh, Jaldish Singh, Jashant Singh, Jennifer Lapidus, Kulwant Singh, Kundan Singh, Maninder Kaur, Michael Kalanty, P. P. Singh, Paramjit Kaur, Rajinder Kaur, Sardar Roop Singh, Tara Jensen, Thom Leonard and Yaakov Horowitz.

OCTOPUS

Devin Bartley and Sylvia Earle.

THANKS

This book was birthed after many false starts—borne out of the inspiration of Stefano Padulosi and Luigi Guarino; through sweat, tears, a flea infestation, a hobbling muscle sprain and a broken heart; with more despair and joy than I could have imagined. It is the result of the enduring faith of Mark Tauber and brother-husband Gideon Weil—and the sustained support, brilliance and creativity of Kim Scherman.

Every person listed has extended more love, wisdom, generosity, patience, kindness and deliciousness than I deserved—particularly Luigi Guarino, Lale Platin, Cristian Melo, Frances Bekele, Vicente Norero, Paul Bordoni, Paul Boettcher, Alok Gupta, Mike Mamo, Moata Raya, Nick Benson, Alice Lieberman, Amy Westerveldt, Deonna Kelli Sayed, Cesare Casella, Prashant Patel, Dan Barber, Annie Wilsey, Carolyn Micek, Jennifer Franck and Jeni Rogers.

I am particularly grateful to Ann Noble, Seneca Klassen, Mark Christian, Timothy Hill, Counter Culture Coffee, Mark Dredge and Michael Kalanty for the generous use of their flavor guide and wheels. And to Caleb Taft, Aaron Wood, Peter Giuliano, Kc Reynolds, Darin Sukha, Naailah Ali, Matthew Escalante, Chloé Doutre-Roussel, Brad Kintzer, Katie Gilmer, Maryuxi Espinoza, Sunita de Tourreil, Randy Mosher, Hande Leimer, Heather Carlucci and all the cuppers at the Ethiopian Coffee Exchange for helping me understand and navigate the sensory experiences of wine, chocolate, coffee, beer and bread.

Finally, thanks to Miles Doyle for editing my words with fondness and care; Suzanne Quist, my production manager, Dianna Stirpe, my copy editor, and Judith Riotto, my proofreader, for being meticulous and compassionate; Caitlin Garing and Therese Plummer for making the printed word come to life in audio form; and Hilary Lawson, Melinda Mullin, Laina Adler, Kim Dayman and everyone at HarperOne who helped make this book possible and shared it with the world.

I am indebted to all of you—for fighting for biodiversity, believing in me and loving the book as much as I do.

INTRODUCTION / WHAT'S AT STAKE

My family at the Adahan Hotel, Alice Julier, Alice Lieberman, Amy Westerveldt, Anna Lappé, Annie Wilsey, Antonello Colonna, Anuradha Mittal, Ardeta Gjikola, Aviva Klein, Braulio Ferreira de Souza Dias, Brenda Ueland, Cade Cruickshank, Carolin Bothe-Tews, Carolyn Micek, Cesare Casella and the Department of Nourishment Arts (DNA), Chris Boswell, Christine Ellem, Colin Khoury, Courtney Southern, Daniel Dermitzel, Daniele Dellorco, David Ladik, David Runia and my family at Queen's College, Deepti Sethi, Defne Koryürek, Deonna Kelli Sayed, Dick Gray, Dizzy and Charlie Howard, Dylan Gray, Emile Frison, Francesa Giampieri, François Mazaud, Gabriella Barbati, Gail Leonard, Gamze Platin, Gary Nabhan, Gauri Salokhe, Gee and Jim Scherman, George Minot, Gideon Weil, Hazel Henderson, Heather Carlucci, Hema Sethi, Irwin Miller, Jai Singh, Jay Gottfried, Jennifer Franck, Jerome Waag, Kat Tan-Comte, Katherine Davis, Katherine Jones, Katie Parla, Kelly Medford, Kim Scherman, Kirsten Khire, Kit Wilkins, Lale Platin, Lauren Lazin, Laurie McClure, Luigi Guarino, Marc Pernick, Maria Garruccio, Matt Hawley, Matt Lehrman, Michelle Wyman, Miles Doyle, Nick Benson, Nolan Kappelman, Patricia Merrikin and the FAO Library, Patrick Dollard and the Center for Discovery, Patty Noland, Paul Boettcher, Paul Bordoni, Paul Hawken, Paul Willis, Phil Morrison, Rebecca Subbiah, Robert Palmer, Sagar Sethi, Savitri Madan, Schuyler Kraus, Sedat Aklan, Shannon McGill, Stefano Padulosi, Tara Bryant, Tim Stinson, TJ Miller, Tom Kostigen, Toshi Singh, Tresa Carter, Varun Mehra, Vasif Kortun, and Yashpal and Damyanti Amarsingh.

WINE

Ann Noble, Bernard Prins, Caleb Taft, Deborah Golino, John Preece, Nancy Sweet, Peter Fanucchi, Scott Schultz, Sébastien Julliard, Snow Barlow and Thierry Lacombe.

CHOCOLATE

Adrian Sieunarine, Adriana Lucas, AGROCALIDAD, Alberto Nácer, ANECACAO, Brad Kintzer, Chloé Doutre-Roussel, Cristian Melo, Darin Sukha, Diego Vizcaino, Frances Bekele, Gino Dalla Gasperina, International Cocoa Organization (ICCO), Jackie Marks, Jeffrey Stern, Joshua Loomes, Katie Gilmer, Kum Kum Bhavnani, Luciana Melo, Mark Christian, Maryuxi Espinoza, Matthew Escalante, Melissa Strabenow, Michael Dorsey, Naailah Ali, Nadia Rosales, Pathmanathan Umaharan and the Cocoa Research Centre, Red Thalhammer, Sarah Khan and the generosity of the TCF Meal by Meal Seed Grant, Seneca Klassen, Sergio and Gloria Cedeño, Sonia Hardy, Sunita de Tourreil, Tad Lombardo and Vicente Norero.

COFFEE

Aaron Wood, Abbie Yunita, Alan Nietlisbach, Alazar Mengstu, Alemayehu Teressa Negawo, Alemitu Tilahun, Allen Leibowitz, Andenet Bekele, Bethelehem Ketema, Brendan Gleeson and the Melbourne Sustainable Society Institute (MSSI), Caitlin McCarthy-Garcia, Christa Lachenmayr, Christopher Mellen, David Roche, Degu Kebede, Eileen Kinney, Elise Pulbrook, Elsa Mehary, Finchwa Farmers Cooperative, Gemedo Dalle Tussie, Hanna Neuschwander, Jennifer Hattam, John Di Ruocco, Kimberly Easson, Maranatha Degu, Mauricio Melara, Mesfin Tekle, Michael Sheridan, Mike Mamo, Moata Raya and TechnoServe, Mohan Singh, Peter Giuliano, Phyllis Johnson, Sisay Tesfaye and researchers at the Jimma Agricultural Research Center (JARC), Stefan Fillipo, Svane Bender-Kaphengst, Tadesse Meskele, Tim Schilling and World Coffee Research, Timothy Hill and Counter Culture Coffee, Tsion Taye, Vanessa Miles, Warren Langley, Werko Robe, Workineh Mulat, Yukro Multipurpose Farmers Cooperative and Zeyede Nigat.

BEER

Andrew Chapple, Ben Ott, Chris Bond, Ian Rycroft, Julie Sprau, Mark Dredge, Mary and Derek Zelmer, National Collection of Yeast Cultures (U.K.), Paige Latham, Prashant Patel, Stan Hieronymus and Steve James.

BREAD

Alok Gupta, Amarjit Bhanwer, Beti Minkin, Dan Barber, Dave Meckler, Devinder Sharma, Eli Rogosa, Gyani Kewal Singh, Heather Carlucci, Jason Taylor, Jindar Bindra, Maneesh Chatrath, Meredith Morse, Michael Kalanty, Nina Virdi, Om Prakash Amarsingh, Sharan Chatrath and Thom Leonard.

OCTOPUS

Terry Sunderland and the Center for International Forestry Research (CIFOR), Lauren Flick, Sylvia Earle, Victor Pisani and WOMADelaide.

ENDNOTES

INTRODUCTION

1. "Tastes Differ—How Taste Preferences Develop," European Food Information Council, last modified February 2011, http://www.eufic.org/article/en/artid/how-taste-preferences-develop/.
2. Wendell Berry, *What Are People For?* (Berkeley, CA: Counterpoint, 2010), 146.
3. Berry, *What Are People For?*, 152.

WHAT'S AT STAKE

1. Bonnie Liebman, "The Changing American Diet: A Report Card," *Nutrition Action Healthletter*, Center for Science in the Public Interest, September 2013, 10–11, http://cspinet.org/new/pdf/changing_american_diet_13.pdf.
2. "What Is Agricultural Biodiversity," Convention on Biological Diversity, accessed August 12, 2014, http://www.cbd.int/agro/whatis.shtml.
3. Ximena Flores Palacios, "Contribution to the Estimation of Countries' Interdependence in the Area of Plant Genetic Resources" (Background Study Paper No. 7 Rev. 1, prepared at the request of the Food and Agriculture Organization of the United Nations [FAO], Commission on Genetic Resources for Food and Agriculture, 1998).
4. "Top 100 Retailers Chart 2014," National Retail Federation, accessed August 12, 2014, https://nrf.com/2014/top100-table.
5. "Varieties," U.S. Apple Association, accessed April 17, 2015, http://www.usapple.org/index.php?option=com_content&view=article&id=21&Itemid=21.
6. Charles Siebert, "Food Ark," *National Geographic*, July 2011, http://ngm.nationalgeographic.com/2011/07/food-ark/siebert-text.
7. Colin K. Khoury et al., "Increasing Homogeneity in Global Food Supplies and the Implications for Food Security," *Proceedings of the National Academy of Sciences of the United States of America (PNAS)* 111, no. 11 (March 2014): 4001–6, doi:10.1073/pnas.1313490111.

8. Colin Khoury, "Will the Pursuit of Food Security Weaken the Resilience of Global Food Systems?" *Agriculture and Ecosystems* (blog), CGIAR Research Program on Water, Land and Ecosystems, May 16, 2014, http://wle.cgiar.org/blogs/2014/05/16/will-pursuit-food-security-weaken-resilience-global-food-systems/.

9. Raj Patel, *Stuffed and Starved: Markets, Power and the Hidden Battle for the World Food System* (New York: Melville House, 2012).

10. Food and Agriculture Organization of the United Nations (FAO), International Fund for Agricultural Development (IFAD), and World Food Programme (WFP), *The State of Food Security in the World 2015: Meeting the 2015 International Hunger Targets: Taking Stock of Uneven Progress* (Rome: FAO, 2015), 1; and Marie Ng et al., "Global, Regional, and National Prevalence of Overweight and Obesity in Children and Adults During 1980–2013: A Systematic Analysis for the Global Burden of Disease Study 2013," *Lancet* 384, no. 9945 (August 2014): 770, http://dx.doi.org/10.1016/S0140-6736(14)60460-8.

11. Veronique Greenwood, "You Are Your Bacteria: How the Gut Microbiome Influences Health," *Time*, August 29, 2013, http://science.time.com/2013/08/29/you-are-your-bacteria-how-the-gut-microbiome-influences-health/.

12. Catherine A. Lozupone et al., "Diversity, Stability and Resilience of the Human Gut Microbiota," *Nature* 489, no. 7415 (September 2012): 220–30, doi:10.1038/nature11550.

13. Food and Agriculture Organization of the United Nations (FAO), *The State of the World's Plant Genetic Resources for Food and Agriculture* (Rome: FAO, 1997), 14, ftp://ftp.fao.org/docrep/fao/meeting/015/w7324e.pdf.

14. "World's Gene Pool Crucial for Survival," Food and Agriculture Organization of the United Nations (FAO), April 15, 2013, http://www.fao.org/news/story/en/item/174330/icode/; and Wilhelm Gruissem, "Global Plant Council: A Coalition of Plant and Crop Societies Across the Globe, Global Needs and Contributions from Plant Science" (presented at the 7th EPSO Conference, Porto Heli, Greece, September 2, 2013).

15. "World's Gene Pool," FAO.

16. B. N. Ekesa, M. K. Walingo, and M. O. A. Onyango, "Role of Agricultural Biodiversity on Dietary Intake and Nutrition Status of Preschool Children in Matungu Division, Western Kenya," *African Journal of Food Science* 2, no. 3 (March 2008): 26–32, http://www.academicjournals.org/journal/AJFS/article-abstract/369BA2F16311.

17. Christy Chamy, "Wheat Rust: The Fungal Disease That Threatens to Destroy the World Crop," *The Independent*, April 20, 2014, http://www.independent.co.uk/news/uk/home-news/wheat-rust-the-fungal-disease-that-threatens-to-destroy-the-world-crop-9271485.html.

18. Veronique Durroux, "Partnerships and Prompt Response, Key to Prevent the Spread of Banana Fusarium Wilt ('Panama Disease')," CGIAR Research Program on Roots, Tubers and Bananas, September 12, 2014, http://www.rtb.cgiar.org/partnerships-and-prompt-response-key-to-prevent-the-spread-of-banana-fusarium-wilt-panama-disease/.

19. Marcello Saitta et al., "Compounds with Antioxidant Properties in Pistachio (*Pistacia vera* L.) Seeds," in *Nuts and Seeds in Health and Disease Prevention*, ed. Victor R. Preedy, Ronald Ross Watson, Vinood B. Patel (London: Academic Press, 2011), 910.

20. Andrew F. Smith, ed., *The Oxford Companion to American Food and Drink* (Oxford: Oxford University Press, 2009), 462.

21. "History," American Pistachio Growers, accessed August 12, 2014, http://american pistachios.org/power-of-pistachios/history.

22. Jimmy Carter, "Executive Order 12211—Sanctions Against Iran" (issued to Congress April 17, 1980).

23. "2014 Pistachio Bearing Acreage, Production and Yield per Acre," The Administrative Committee for Pistachios, annual industry statistics prepared January 2015, http://www.acpistachios.org/pdf/2014Statistics.pdf; and Craig Kallsen, "Pistachio Breeding: A Lifetime May Not Be Long Enough," *Western Farm Press*, May 20, 2003, http://westernfarmpress.com/pistachio-breeding-lifetime-may-not-be-long-enough.

24. "Food Insecurity," Child Trends Databank, last modified December 2014, http://www.childtrends.org/?indicators=food-insecurity.

25. Eric Holt-Giménez, "The World Food Crisis: What Is Behind It and What We Can Do," *Hunger Notes*, World Hunger Education Service, October 23, 2008, http://www.worldhunger.org/articles/09/editorials/holt-gimenez.htm.

26. Eric Holt-Giménez, "We Already Grow Enough Food for 10 Billion People—and Still Can't End Hunger," *Huffington Post*, May 2, 2012, http://www.huffingtonpost.com/eric-holt-gimenez/world-hunger_b_1463429.html.

27. "Smallholder Farming and Achieving Our Development Goals," Landesa Rural Development Institute, July 2014, 1, http://www.landesa.org/wp-content/uploads/Issue-Brief-Smallholder-Farming-and-Achieving-Our-Development-Goals.pdf; and "Hungry for Land: Small Farmers Feed the World with Less Than a Quarter of All Farmland," GRAIN, May 28, 2014, https://www.grain.org/article/entries/4929-hungry-for-land-small-farmers-feed-the-world-with-less-than-a-quarter-of-all-farmland.

28. M. Ann Tutwiler, "Celebrating Smallholder Farmers and Rural Women—DG Dialogues," Bioversity International, March 10, 2014, http://www.bioversityinternational.org/news/detail/celebrating-smallholder-farmers-and-rural-women-dg-dialogues/.

29. Mona Chalabi and John Burn-Murdoch, "McDonald's 34,492 Restaurants: Where Are They?" *The Guardian*, July 17, 2013, http://www.theguardian.com/news/datablog/2013/jul/17/mcdonalds-restaurants-where-are-they#data.

30. D.H. and L.D., "The Big Mac Index: Global Exchange Rates, To Go," *The Economist*, January 22, 2015, http://www.economist.com/content/big-mac-index.

31. Jack Linshi, "This Is McDonald's Big Plan to Win You Over," *Time*, October 22, 2014, http://time.com/3531339/mcdonalds-earnings-regional-personal-menu/.

32. Michael Moss, *Salt Sugar Fat: How the Food Giants Hooked Us* (New York: Random House, 2014).

33. "McDonald's USA Nutrition Facts for Popular Menu Items," McDonald's Corporation, accessed June 23, 2015, http://nutrition.mcdonalds.com/getnutrition/nutritionfacts.pdf; and Roberto A. Ferdman, "There Are 19 Ingredients in McDonald's French Fries," *Wonkblog, Washington Post*, January 22, 2015, http://www.washingtonpost.com/blogs/wonkblog/wp/2015/01/22/there-are-19-ingredients-in-mcdonalds-french-fries/.

34. Rudd Center for Food Policy and Obesity, Yale University, "Food & Addiction" (paper presented at the Conference on Eating and Dependence, New Haven, Connecticut, July 2007).

35. S. Jay Olshansky et al., "A Potential Decline in Life Expectancy in the United States in the 21st Century," *New England Journal of Medicine* 352 (March 2005): 1138-45, doi:10.1056/NEJMsr043743.

36. "Expenditures on Food and Alcoholic Beverages That Were Consumed at Home by Selected Countries," USDA Economic Research Service, accessed June 23, 2015, http://www.ers.usda.gov/data-products/food-expenditures.aspx#26654.

37. "100 Years of U.S. Consumer Spending: Data for the Nation, New York City, and Boston: 1934-36," U.S. Department of Labor, Bureau of Labor Statistics, last modified August 3, 2006, 16, http://www.bls.gov/opub/uscs/1934-36.pdf.

38. Johannes Frasnelli, "The Perception of Flavor—Retronasal Olfaction," *Odor Management* (blog), Odotech, February 28, 2012, http://blog.odotech.com/perception-flavor-retronasal-olfaction.

39. Helen Fields, "Fragrant Flashbacks," *Observer* 25, no. 4 (April 2012), http://www.psychologicalscience.org/index.php/publications/observer/2012/april-12/fragrant-flashbacks.html.

40. Diane Ackerman, *A Natural History of the Senses* (New York: Vintage Books, 1995), 5.

41. Carlo Agostoni et al., "Free Amino Acid Content in Standard Infant Formulas: Comparison with Human Milk," *Journal of the American College of Nutrition* 19, no. 4 (September 2000): 434-38, doi:10.1080/07315724.2000.10718943.

42. Charles Spence and Mary Kim Ngo, "Assessing the Shape Symbolism of the Taste, Flavour, and Texture of Foods and Beverages," *Flavour* 1 (July 2012), doi:10.1186/2044-7248-1-12.

43. Russ Jones, "Bittersweet Symphony," *Condiment Junkie* (blog), March 31, 2014, http://condimentjunkie.co.uk/blog/2015/4/27/bittersweet-symphony.

44. Pierre Bourdieu, *Distinction: A Social Critique of the Judgement of Taste* (Cambridge, MA: Harvard University Press, 1987).

45. Daniel Luzer, "How Lobster Got Fancy," *Pacific Standard*, June 7, 2013, http://www.psmag.com/business-economics/how-lobster-got-fancy-59440.

46. Mandy Oaklander, "Health Food Face-Off: Kale vs. Spinach," *Prevention*, accessed June 23, 2015, http://www.prevention.com/content/whats-healthier-kale-or-spinach.

47. Lara A. Latimer, Lizzy Pope and Brian Wansink, "Food Neophiles: Profiling the Adventurous Eater," *Obesity* 23 (2015), doi:10.1002/oby.21154.

48. R. Loewen and P. Pliner, "Effects of Prior Exposure to Palatable and Unpalatable Novel Foods on Children's Willingness to Taste Other Novel Foods," *Appetite* 32, no. 3 (June 1999): 351-66, doi:10.1006/appe.1998.0216.

49. Patricia Pliner, Marcia Pelchat, and Marius Grabski, "Reduction of Neophobia in Humans by Exposure to Novel Foods," *Appetite* 20, no. 2 (April 1993): 111-23, doi:10.1006/appe.1993.1013.

50. Louise Erdrich, *The Painted Drum: A Novel (P.S.)* (New York: Harper Perennial, 2005), 274.

WINE

1. Paul Rozin, "Preadaptation and the Puzzles and Properties of Pleasure," in *Well Being: The Foundations of Hedonic Psychology*, ed. Daniel Kahneman, Edward Diener, and Norbert Schwarz (New York: Russell Sage, 1999), 112.
2. Clinton B. Wright et al., "Reported Alcohol Consumption and Cognitive Decline: The Northern Manhattan Study," *Neuroepidemiology* 27, no. 4 (December 2006): 201–7, doi:10.1159/000096300; and Timo E. Strandberg et al., "Alcoholic Beverage Preference, 29-Year Mortality, and Quality of Life in Men in Old Age," *Journals of Gerontology* 62, no. 2 (2007): 213–18, http://biomedgerontology.oxfordjournals.org /content/62/2/213.full.pdf+html.
3. Michael Adrian Peters, "In Vino Veritas: In Wine the Truth," *Journal of Aesthetic Education* 45, no. 3 (January 2011): 114–17, doi:10.1353/jae.2011.0028.
4. Lalli Nykänen and Heikki Suomalainen, *Aroma of Beer, Wine and Distilled Alcoholic Beverages* (Hingham, MA: Kluwer Academic Publishers, 1983), V.
5. Randy Mosher, *Tasting Beer: An Insider's Guide to the World's Greatest Drink* (North Adams, MA: Storey Publishing, 2009), 29.
6. Angus L. Hughson and Robert A. Boakes, "The Knowing Nose: The Role of Knowledge in Wine Expertise," *Food Quality and Preference* 13, no. 7 (October 2002): 463–72.
7. Sally Wiggins, "Talking About Taste: Using a Discursive Psychological Approach to Examine Challenges to Food Evaluations," *Appetite* 43, no. 1 (September 2004): 29, doi:10.1016/j.appet.2004.01.007.
8. "Sniff Study Suggests Humans Can Distinguish More Than 1 Trillion Scents," Rockefeller University Newswire, March 20, 2014, http://newswire.rockefeller .edu/2014/03/20/sniff-study-suggests-humans-can-distinguish-more-than-1-trillion- scents/.
9. C. Bushdid et al., "Humans Can Discriminate More Than 1 Trillion Olfactory Stimuli," *Science* 343, no. 6177 (March 2014): 1370–72, doi:10.1126/science.1249168.
10. Diane Ackerman, *A Natural History of the Senses* (New York: Vintage Books, 1995), 20.
11. Ackerman, *A Natural History of the Senses*, 6.
12. Richard Axel, "The Molecular Logic of Smell," *Scientific American*, September 2006, 72, doi:10.1038/scientificamerican0906-68sp.
13. Alex Stone, "Smell Turns Up in Unexpected Places," *New York Times*, October 13, 2014, http://www.nytimes.com/2014/10/14/science/smell-turns-up-in-unexpected- places.html?smid=tw-nytimesscience&_r=0.
14. Robert Wighton Moncrieff, "What Is Odor? A New Theory," *American Perfumer* 54 (1949): 453.
15. Carl Zimmer, "The Smell of Evolution," *National Geographic*, December 11, 2013, http://phenomena.nationalgeographic.com/2013/12/11/the-smell-of-evolution/.
16. Asifa Majid and Niclas Burenhult, "Odors Are Expressible in Language, as Long as You Speak the Right Language," *Cognition* 130, no. 2 (February 2014): 266–70, doi:10.1016/j.cognition.2013.11.004.

17. S. Julliard et al., "Prospection et Valorisation du Cépage Trousseau Gris dans les Charentes [Prospecting and Exploitation of Varietal Trousseau Gris in Charentes]," *Revue des Oenologues et des Techniques Vitivinicoles et Oenologoques* 122 (2007): 46–48.

18. "Classification of Plants," The Seed Site, accessed May 20, 2015, http://theseedsite.co.uk/class.html.

19. "California Grape Acreage Report, 2013 Summary," California Department of Food and Agriculture, April 15, 2014, http://www.nass.usda.gov/Statistics_by_State/California/Publications/Fruits_and_Nuts/201403grpac.pdf.

20. "Meet the Grapes," California Wines, Wine Institute of California, accessed December 20, 2014, http://www.discovercaliforniawines.com/meet-the-grapes/.

21. "Mission," Professional Friends of Wine, last modified October 29, 2011, http://www.winepros.org/wine101/grape_profiles/mission.htm.

22. Luke 7:33–34 (New King James Version).

23. "California Marks 150th Anniversary of Gold Discovery: An Era That Transformed the State's Wine Industry," Wine Institute, May 1, 2006, http://www.wineinstitute.org/resources/pressroom/05012006.

24. "History of Phylloxera," Calwineries, accessed December 10, 2014, http://www.calwineries.com/learn/grape-growing/pests-and-diseases/history-of-phylloxera.

25. "Complete California Wine History from the Early 1800s to Today," The Wine Cellar Insider, accessed December 12, 2014, http://www.thewinecellarinsider.com/california-wine/california-wine-history-from-early-plantings-in-1800s-to-today/#ixzz3LoqHJhNM.

26. Christopher Bland, "When France Withered on the Vine," *The Telegraph*, March 8, 2004, http://www.telegraph.co.uk/culture/books/3613731/When-France-withered-on-the-vine.html.

27. "Crop Wild Relatives," USDA Forest Service, accessed December 12, 2014, http://www.fs.fed.us/wildflowers/ethnobotany/wildrelatives.shtml.

28. Bland, "When France Withered on the Vine."

29. "How to Manage Pests: UC Pest Management Guidelines: Grape," University of California Agriculture & Natural Resources, last modified April 2014, http://www.ipm.ucdavis.edu/PMG/r302300811.html.

30. Karen MacNeil, *The Wine Bible* (New York: Workman Publishing, 2001), 636–43.

31. "The 1976 Paris Tasting," Stag's Leap Wine Cellars, accessed December 14, 2014, https://www.cask23.com/history/parisTasting/.

32. Thane Peterson, "The Day California Wines Came of Age," *Bloomberg Business*, May 7, 2001, http://www.bloomberg.com/bw/stories/2001-05-07/the-day-california-wines-came-of-age.

33. Alan Hamilton and David Sanderson, "California Reds Win by a Nose in Tasting Rematch," *The Times*, last modified May 25, 2006, http://www.thetimes.co.uk/tto/news/world/europe/article2602209.ece.

34. Cyril Penn, "WBM 30," *Wine Business Monthly*, February 2013, http://www.winebusiness.com/wbm/?go=getArticleSignIn&dataId=111054.

35. Alice Feiring, "That Wine Isn't Vegan and Other Reasons to Go Natural," *Civil Eats*, December 16, 2014, http://civileats.com/2014/12/16/that-wine-isnt-vegan-and-other-reasons-to-go-natural/.

36. Kym Anderson and Nanda Aryal, "Where in the World Are Various Winegrape Varieties Grown? Evidence from a New Database," Wine Economics Research Centre, University of Adelaide, December 2013, 9, http://www.academia.edu/6688711/0213-where-are-wine-grape-varieties-grown-dec2013.

37. *Neglected and Underutilized Plant Species: Strategic Action Plan of the International Plant Genetic Resources Institute* (Rome: IPGRI, 2002), 12, https://books.google.com.my/books?id=-Jy6A-K8eMcC&source=gbs_navlinks_s.

38. "Major Crops Grown in the United States," U.S. Environmental Protection Agency, last modified April 11, 2013, http://www.epa.gov/agriculture/ag101/cropmajor.html.

39. Marti, "What Is the Difference Between Soil and Dirt?" Big How, accessed December 15, 2014, http://bighow.org/1895278-What_Is_the_Difference_Between_Soil__amp__Dirt_.html.

40. "Spotlighting Humanity's 'Silent Ally,' UN Launches 2015 International Year of Soils," *UN News Centre*, December 5, 2014, http://www.un.org/apps/news/story.asp?NewsID=49520#.VYl4eFXBzGe.

41. "Nothing Dirty Here: FAO Kicks Off International Year of Soils 2015," Food and Agriculture Organization of the United Nations (FAO), December 4, 2014, http://www.fao.org/news/story/en/item/270812/icode/.

42. Heinrich Böll Stiftung and Institute for Advanced Sustainability Studies, *Soil Atlas 2015: Facts and Figures About Earth, Land and Fields* (Berlin: Herinrich Böll Stiftung and IASS, 2015), 7.

43. "Global Plans of Action Endorsed to Halt the Escalating Degradation of Soils," Food and Agriculture Organization of the United Nations (FAO), July 24, 2014, http://www.fao.org/news/story/en/item/239341/icode/.

44. "The Skin of the Earth," Forces of Change, Smithsonian Institution, accessed June 23, 2015, http://forces.si.edu/soils/02_01_01.html.

45. "What on Earth Is Soil?" Gulf of Mexico Program, U.S. Environmental Protection Agency, last modified January 30, 2014, http://epa.gov/gmpo/edresources/soil.html.

46. Becca Yeamans-Irwin, "Evolution of Soil Over the Past 1 Million Years Determines Terroir in Wine," *The Academic Wino* (blog), May 22, 2013, http://www.academicwino.com/2013/05/evolution-of-soil-determines-terroir-in-wine.html/.

47. Rainer Maria Rilke, *Sonnets to Orpheus* (Hanover, NH: Wesleyan University Press: 1987), 29.

48. Calanit Bar-Am and Daniel A. Sumner, "Economic Effects of Climate Change on California Wine Industry—Research in Progress," University of California Agricultural Issue Center, accessed June 23, 2015, http://aic.ucdavis.edu/publications/posters/wine_climate6.pdf; and Lee Hannah et al., "Climate Change, Wine, and Conservation," *Proceedings of the National Academy of Sciences of the United States of America (PNAS)* 110, no. 17 (April 2013): 6907–12, doi:10.1073/pnas.1210127110.

49. Haya El Nasser, "Raise a Glass: Drought Not Disrupting California Wine (Yet)," *Al Jazeera America*, April 9, 2014, http://america.aljazeera.com/articles/2014/4/9/drought-californiawine.html.

50. "California Agricultural Production Statistics," California Department of Food and Agriculture, accessed June 23, 2015, http://www.cdfa.ca.gov/statistics/.

51. Richard Howitt et al., "Economic Analysis of the 2014 Drought for California Agriculture," Center for Watershed Services, University of California, Davis, July 23, 2014, https://watershed.ucdavis.edu/files/biblio/DroughtReport_23July2014_0.pdf.

52. Christopher B. Field et al., *Climate Change 2014: Impacts, Adaptation, and Vulnerability: Summary for Policymakers* (report presented at the Intergovernmental Panel on Climate Change, March 31, 2014), 17, https://ipcc-wg2.gov/AR5/images/uploads/IPCC_WG2AR5_SPM_Approved.pdf.

53. Food and Agriculture Organization of the United Nations (FAO), "The State of Ex Situ Conservation," in *The Second Report on the State of the World's Plant Genetic Resources for Food and Agriculture (PGRFA)* (Rome: FAO, 2010), 58.

54. "Yearbook Tables: Table A-17—Utilized Production of Selected Fruit and Tree Nuts in the United States, by State, 2012," USDA Economic Research Service, last modified October 31, 2014, http://www.ers.usda.gov/data-products/fruit-and-tree-nut-data/yearbook-tables.aspx#40788.

55. Stephanie Greene, "National Inventory Takes Stock of Crops' Wild Relatives," *AgResearch Magazine*, January 2014, http://agresearchmag.ars.usda.gov/2014/jan/crops.

56. "Corn and Its Untamed Cousins: Wild Genes in Domestic Crops," Understanding Evolution, University of California, Berkeley, accessed December 15, 2014, http://evolution.berkeley.edu/evolibrary/article/agriculture_03.

57. Barry Estabrook, "Why Is This Wild, Pea-Sized Tomato So Important?" Smithsonian.com, July 22, 2015, http://www.smithsonianmag.com/travel/why-wild-tiny-pimp-tomato-so-important-180955911/?no-ist.

58. Suzanne Goldenberg, "The Doomsday Vault: The Seeds That Could Save a Post-apocalyptic World," *The Guardian*, May 20, 2015, http://www.theguardian.com/science/2015/may/20/the-doomsday-vault-seeds-save-post-apocalyptic-world.

59. Patrick Matthews, *Real Wine: The Rediscovery of Natural Winemaking* (London: Mitchell Beazley, 2000), 12–13.

60. Sandor Ellix Katz, *The Art of Fermentation* (White River Junction, VT: Chelsea Green Publishing, 2012), xviii, xix.

61. Joel D. Mainland et al., "The Missense of Smell: Functional Variability in the Human Odorant Receptor Repertoire," *Nature Neuroscience* 17 (December 2013): 114–20, doi:10.1038/nn.3598.

62. Ellen Byron, "Uncork the Nose's Secret Powers," *Wall Street Journal*, February 20, 2013, http://www.wsj.com/articles/SB10001424127887323696404578300182010199640.

63. Gil Morrot, Frédéric Brochet, and Denis Dubourdieu, "The Color of Odors," *Brain and Language* 79, no. 2 (August 2001): 309–20, doi:10.1006/brln.2001.2493.

64. Frédéric Brochet, "Chemical Object Representation in the Field of Consciousness" (application presented for the grand prix of the Académie Amorim following work

carried out toward a doctorate from the Faculty of Oenology, General Oenology Laboratory, 2001).

65. Hilke Plassmann et al., "Marketing Actions Can Modulate Neural Representations of Experienced Pleasantness," *Proceedings of the National Academy of Sciences of the United States of America (PNAS)* 105, no. 3 (January 2008): 1050–54, doi:10.1073/pnas.0706929105.

66. Rilke, *Sonnets to Orpheus,* 29.

67. George M. Taber, *Judgment of Paris: California vs. France and the Historic 1976 Paris Tasting That Revolutionized Wine* (New York: Scribner, 2005), 165.

CHOCOLATE

1. L. H. Yao et al., "Flavonoids in Food and Their Health Benefits," *Plant Foods for Human Nutrition* 59, no. 3 (Summer 2004): 113–22, doi:10.1007/s11130-004-0049-7.

2. Bianca Fuhrman and Michael Aviram, "Flavonoids Protect LDL from Oxidation and Attenuate Atherosclerosis," *Current Opinion in Lipidology* 12, no. 1 (February 2001): 41–48, http://journals.lww.com/co-lipidology/Abstract/2001/02000/Flavonoids_protect_LDL_from_oxidation_and.8.aspx.

3. "Health Benefits of Cocoa Flavonoids," European Food Information Council (EUFIC), last modified July 2006, http://www.eufic.org/article/en/artid/health-benefits-cocoa-flavanoids/.

4. "Heart Health Benefits of Chocolate," Cleveland Clinic, last modified January 2012, http://my.clevelandclinic.org/services/heart/prevention/nutrition/food-choices/benefits-of-chocolate.

5. Anne Marie Helmenstine, "Theobromine Chemistry," About Education, About.com, accessed August 12, 2014, http://chemistry.about.com/od/factsstructures/a/theobromine-chemistry.htm.

6. Pamela Zurer, "Chocolate May Mimic Marijuana in Brain," *Chemical & Engineering News* 74, no. 36 (September 1996): 31–32, doi:10.1021/cen-v074n036.p031a.

7. Diane Ackerman, *A Natural History of the Senses* (New York: Vintage Books, 1995), 156.

8. "The Chocolate League Tables 2014: Top 20 Consuming Nations," Euromonitor International (compiled by ConfectionaryNews.com), last modified October 2014, http://www.targetmap.com/viewer.aspx?reportId=38038.

9. Oliver Nieburg, "The Chocolate League Tables 2014: Top 20 Consuming Nations," Confectionery News.com, October 9, 2014, http://www.confectionerynews.com/Markets/Chocolate-consumption-by-country-2014.

10. Sophie D. Coe and Michael D. Coe, *The True History of Chocolate* (New York: Thames and Hudson, 2000), 61.

11. Amanda Fiegl, "A Brief History of Chocolate," Smithsonian.com, March 1, 2008, http://www.smithsonianmag.com/arts-culture/a-brief-history-of-chocolate-21860917/?no-ist.

12. L. V. Anderson, "Cuckoo for Cocoa: Are Women Really Crazy About Chocolate?" *Slate,* February 13, 2012, http://www.slate.com/articles/double_x/doublex/2012/02/valentine_s_day_do_women_crave_chocolate_or_is_that_a_stereotype_.html.

13. Mindy Badia and Bonnie L. Gasior, *Crosscurrents: Transatlantic Perspectives on Early Modern Hispanic Drama* (Lewisburg, PA: Bucknell University Press, 2006), 29.

14. Fiegl, "A Brief History of Chocolate."

15. Louis E. Grivetti and Howard-Yana Shapiro, *Chocolate: History, Culture, and Heritage* (Hoboken, NJ: John Wiley & Sons, 2009), 5.

16. Coe and Coe, *The True History of Chocolate*, 108.

17. Coe and Coe, *The True History of Chocolate*, 126.

18. "Sweet News for Sweet Lovers," Chocolate.org, accessed February 20, 2015, http://www.chocolate.org/articles/sweet-news-for-sweet-lovers.html; and Tori Avey, "Learn Why These 10 Foods Are Edible Aphrodisiacs," Food: The History Kitchen, PBS, February 10, 2014, http://www.pbs.org/food/the-history-kitchen/10-edible-aphrodisiacs/.

19. Erik Strand, "Tryptophan: What Does It Do?" *Psychology Today*, last modified January 23, 2015, https://www.psychologytoday.com/articles/200309/tryptophan-what-does-it-do.

20. Anahad O'Connor, "The Claim: Chocolate Is an Aphrodisiac," *New York Times*, July 18, 2006, http://www.nytimes.com/2006/07/18/health/18real.html?_r=0.

21. Ackerman, *A Natural History of the Senses*, 154.

22. Ernest Entwistle Cheesman, "Notes on the Nomenclature, Classification and Possible Relationships of Cocoa Populations," *Tropical Agriculture* 21 (1944): 144–59.

23. Juan C. Motamayor et al., "Cacao Domestication I: The Origin of the Cacao Cultivated by the Mayas," *Heredity* 89, no. 5 (November 2002): 380–86, doi:10.1038/sj.hdy.6800156.

24. Charles R. Clement et al., "Origin and Domestication of Native Amazonian Crops," *Diversity* 2, no. 1 (January 2010): 78, doi:10.3390/d2010072.

25. Juan C. Motamayor et al., "Geographic and Genetic Population Differentiation of the Amazonian Chocolate Tree (*Theobroma Cacao* L.)," *PLOS ONE* 3, no. 10 (October 2008): e3311, 2, doi:10.1371/journal.pone.0003311.

26. Jonny O'Callaghan, "Why Does Milk Turn Sour?" *How It Works*, January 20, 2012, http://www.howitworksdaily.com/why-does-milk-turn-sour/.

27. Rosane F. Schwan and Alan E. Wheals, "The Microbiology of Cocoa Fermentation and Its Role in Chocolate Quality," *Critical Reviews in Food Science and Nutrition* 44, no. 4 (2004): 205–21, doi:10.1080/10408690490464104.

28. G. A. R. Wood and R. A. Lass, *Cocoa (Tropical Agriculture)* (Oxford: Wiley-Blackwell, 2001), 513.

29. Josh Cohen, "Fermentation for Dummies," *Tasting Table*, April 21, 2015, http://www.tastingtable.com/cook/national/how-to-ferment-how-to-make-sauerkraut-sandor-katz#ixzz3Z5qMfkk8.

30. Sara Lewis, "The Loves and Lies of Fireflies," TED video, 13:01, filmed and posted March 2014, http://www.ted.com/talks/sara_lewis_the_loves_and_lies_of_fireflies?language=en.

31. "Pollinators—Chocolate Midge," National Park Service, U.S. Department of the Interior, last modified June 20, 2015, http://www.nps.gov/articles/chocolate-midge.htm.

32. "Pollinators—Chocolate Midge."

33. Wood and Lass, *Cocoa (Tropical Agriculture)*, 20.

34. "Growing Cocoa Beans," World Agroforestry Centre, accessed April 17, 2015, http://worldagroforestry.org/treesandmarkets/inaforesta/documents/agrof_cons _biodiv/Ch.3-Growing-Cocoa-Beans.pdf.

35. "Cocoa Market Update," World Cocoa Foundation, last modified April 1, 2014, http://worldcocoafoundation.org/wp-content/uploads/Cocoa-Market-Update-as-of-4-1-2014.pdf.

36. "Fairtrade and Cocoa," Fairtrade Foundation, August 2011, 2, http://www.fairtrade .net/fileadmin/user_upload/content/2009/resources/2011_Fairtrade_and_cocoa _briefing.pdf.

37. "Fairtrade and Cocoa," 2, 3, 6.

38. Corby Kummer, "A Chocolate Maker's Big Innovation," *MIT Technology Review*, June 18, 2013, http://www.technologyreview.com/review/516206/a-chocolate-makers-big-innovation/.

39. "Fine or Flavour Cocoa," International Cocoa Organization, last modified April 13, 2015, http://www.icco.org/about-cocoa/fine-or-flavour-cocoa.html.

40. "Cacao Percentages," The Story of Chocolate, The National Confectioners Association's Chocolate Council, accessed February 20, 2015, http://www.thestoryofchoco late.com/Savor/content.cfm?ItemNumber=3454&navItemNumber=3376.

41. Robert L. Wolke, "Chocolate by the Numbers," *Washington Post*, June 9, 2004, http://www.washingtonpost.com/wp-dyn/articles/A24276-2004Jun8.html.

42. Wolke, "Chocolate by the Numbers."

43. "HERSHEY'S Milk Chocolate Bar," The Hershey Company, accessed February 21, 2015, http://www.thehersheycompany.com/brands/hersheys-bars/milk-chocolate.aspx.

44. "SeriousMilk™ 'Classic' 39% Cacao 58g Bar," TCHO, accessed February 21, 2015, http://www.tcho.com/shop/seriousmilk-classic-milk-chocolate-39-percent-cacao-58g-bar-11221#.

45. "Flavor Focus," TCHO, accessed June 26, 2015, http://www.tcho.com/tchois/flavor-focus/.

46. "List of Goods Produced by Child Labor or Forced Labor," U.S. Department of Labor, 2014, accessed April 17, 2015, http://www.dol.gov/ilab/reports/child-labor /list-of-goods/.

47. Joel Weber, "To Save Chocolate, Scientists Develop New Breeds of Cacao," *Bloomberg Business*, November 13, 2014, http://www.bloomberg.com/news/articles/2014-11-14 /to-save-chocolate-scientists-develop-new-breeds-of-cacao.

48. "The Chocolate Industry," International Cocoa Organization, last modified January 23, 2015, http://www.icco.org/about-cocoa/chocolate-industry.html.

49. Darin A. Sukha et al., "The Impact of Processing Location and Growing Environment on Flavor in Cocoa (*Theobroma Cacao* L.)—Implications for 'Terroir' and Certification—Processing Location Study," *ISHS Acta Horticulturae* 1047 (2014): 255–62, http://www .actahort.org/books/1047/1047_31.htm.

50. Filippo Tommaso Marinetti, *Critical Writings: New Edition*, ed. Günter Berghaus, trans. Doug Thompson (New York: Farrar, Straus and Giroux, 2006), 395.

51. Michael Pollan, *The Botany of Desire: A Plant's-Eye View of the World* (New York: Random House, 2001).

52. Allen M. Young, Marilyn Schaller, and Melanie Strand, "Floral Nectaries and Trichomes in Relation to Pollination in Some Species of Theobroma and Herrania (Sterculiaceae)," *American Journal of Botany* 71, no. 4 (April 1984): 466–80, http://www.jstor.org/stable/2443322.

53. Nancy Miorelli, "The Budding Relationship of a Midge and the Chocolate Flower," Ask an Entomologist, January 7, 2015, https://askentomologists.wordpress.com/2015/01/07/the-budding-relationship-of-a-midge-and-the-chocolate-flower/.

54. Mark Christian, "Bar Chocolate Review: Cacao Butter by Navitas," *C-Spot*, December 15, 2014.

55. David J. Linden, "Food, Pleasure and Evolution," *The Compass of Pleasure* (blog), *Psychology Today*, March 30, 2011, https://www.psychologytoday.com/blog/the-compass-pleasure/201103/food-pleasure-and-evolution.

56. Linden, "Food, Pleasure and Evolution."

57. Nicholas P. Cheremisinoff, *Dust Explosion and Fire Prevention Handbook: A Guide to Good Industry Practices* (Hoboken, NJ: John Wiley & Sons, 2014), 4.

58. Stephen T. Beckett, *The Science of Chocolate* (Cambridge: The Royal Society of Chemistry, 2008), 66.

59. http://www.outsideonline.com/1885131/heart-dark-chocolate.

60. Julia Calderone and Ben Fogelson, "Fact or Fiction?: The Tongue Is the Strongest Muscle in the Body," *Scientific American*, August 15, 2014, http://www.scientificamerican.com/article/fact-or-fiction-the-tongue-is-the-strongest-muscle-in-the-body/.

61. Bijal P. Trivedi, "Neuroscience: Hardwired for Taste," *Nature* 486 (June 2012): S9, doi:10.1038/486S7a; and Liz Day, "Taste Buds Found in Lungs," *Discovery News*, October 26, 2010, http://news.discovery.com/human/health/taste-buds-in-lungs-detect-bitters-could-ease-asthma.htm.

62. "Growth of Taste and Smell Receptors," What's New, Taste and Smell Clinic, last modified November 2007, http://www.tasteandsmell.com/nov07.htm.

63. "How Does Our Sense of Taste Work?" PubMed Health, National Center for Biotechnology Information, last modified January 6, 2012, http://www.ncbi.nlm.nih.gov/pubmedhealth/PMH0072592/.

64. Russell S. J. Keast and Andrew Costanzo, "Is Fat the Sixth Taste Primary? Evidence and Implications," *Flavour* 4 (February 2015), doi:10.1186/2044-7248-4-5.

65. Dipayan Biswas et al., "Something to Chew On: The Effects of Oral Haptics on Mastication, Orosensory Perception, and Calorie Estimation," *Journal of Consumer Research* 41, no. 2 (August 2014): 261–73, doi:10.1086/675739.

66. Marta Yanina Pepino et al., "The Fatty Acid Translocase Gene, CD36, and Lingual Lipase Influence Oral Sensitivity to Fat in Obese Subjects," *Journal of Lipid Research* 53 (March 2012): 564, doi:10.1194/jlr.M021873.

67. "How Does Our Sense of Taste Work?"

68. Jeremy M. Berg, John L. Tymoczko, and Lubert Stryer, "Taste Is a Combination of Senses That Function by Different Mechanisms," in *Biochemistry*, 5th ed. (New York: W. H. Freeman, 2002).

69. George Vierra, "Physiology of Odor and Flavor Perception," Napa Valley College, December 2013, 1, http://www.napavalley.edu/people/gvierra/Documents/Sensory _Evaluation_of_Wine/Smelling%20and%20Tasting%20-%20Physiology.pdf.

70. "How Does Our Sense of Taste Work?"

71. Linda Bartoshuk, "Separate Worlds of Taste," *Psychology Today* 14 (1980): 48–49, 51, 54–56, 63.

72. "Can Cheap Wine Taste Great? Brain Imaging and Marketing Placebo Effects," *Journal of Marketing Research* (press release), April 29, 2015, https://www.ama.org/pub lications/JournalOfMarketingResearch/Documents/pr-jmr.13.0613-brain-imaging-wine-placebo.pdf.

73. "Can Cheap Wine Taste Great?"

74. Rey Gastón Loor et al., "Genetic Diversity and Possible Origin of the Nacional Cacao Type from Ecuador" (paper presented at the International Cocoa Research Conference, San José, Costa Rica, October 9–14, 2006).

75. Cristian J. Melo, "Left Behind: A Farmer's Fate in the Age of Sustainable Development," *Florida International University Electronic Theses and Dissertations*, paper 331, November 8, 2010, 128, http://digitalcommons.fiu.edu/cgi/viewcontent.cgi?article =1378&context=etd.

76. Yaw Adu-Ampomah, "Overview of Cocoa Research Institute of Ghana's Cocoa Reseach Programs" (presentation, Ghana Cocoa Board, June 14, 2013).

77. David Guest, "Black Pod: Diverse Pathogens with a Global Impact on Cocoa Yield," *Phytopathology* 97, no. 12 (December 2007): 1650, doi:10.1094/PHYTO-97-12-1650.

78. "Phytophthora Basics," Forest Phytophthoras of the World, accessed March 3, 2015, http://forestphytophthoras.org/phytophthora-basics.

79. Committee on the Strategic Planning for the Florida Citrus Industry: Addressing Citrus Greening Disease (Huanglongbing) et al., "Appendix L: Witches' Broom Disease Outbreak in Brazil and Control Attempts: Success and Failure in Bahia, Brazil (1989–2009)," in *Strategic Planning for the Florida Citrus Industry: Addressing Citrus Greening* (Washington, DC: National Academies Press, 2010), 305; and Shawn Smallman, "Witches' Broom: The Mystery of Chocolate and Bioterrorism in Brazil," *Introduction to International and Global Studies Blog*, last modified June 6, 2014, http:// introtoglobalstudies.com/2012/03/witches-broom-the-mystery-of-chocolate-and-bioterrorism-in-brazil/.

80. Polya Lesova, "One Cocoa Bean at a Time," *MarketWatch*, April 10, 2008, http:// www.marketwatch.com/story/state-of-bahia-struggles-to-reclaim-brazils-cocoa-bean-glory.

81. "Moniliophthora Roreri (Frosty Pod Rot)," CABI Invasive Species Compendium, last modified December 16, 2014, http://www.cabi.org/isc/datasheet/34779.

82. "Pests & Diseases," International Cocoa Organization, last modified April 10, 2015, http://www.icco.org/about-cocoa/pest-a-diseases.html.

83. "Moniliophthora Roreri (Frosty Pod Rot)."

84. Edward J. Boza et al., "Genetic Characterization of the Cacao Cultivar CCN 51: Its Impact and Significance on Global Cacao Improvement and Production," *Journal of the American Society for Horticulture Science* 139, no. 2 (March 2014): 219–29; and

G. A. R. Wood and R. A. Lass, *Cocoa (Tropical Agriculture)* (Oxford: Wiley-Blackwell, 2001), 84.

85. Steve Striffler, *In the Shadows of State and Capital: The United Fruit Company, Popular Struggle, and Agrarian Restructuring in Ecuador, 1900–1995* (Durham, NC: Duke University Press, 2001), 42.

86. "What Is Panama Disease?" Panama Disease, Wageningen UR, accessed March 6, 2015, http://panamadisease.org/en/theproblem.

87. "Growing Cocoa: Origins of Cocoa and Its Spread Around the World," International Cocoa Organization, last modified March 26, 2013, http://www.icco.org/about-cocoa/growing-cocoa.html.

88. "English—2004/2005 Annual Report," International Cocoa Organization, accessed March 7, 2015, 10, http://www.icco.org/about-us/international-cocoa-agreements /cat_view/1-annual-report/23-icco-annual-report-in-english.html.

89. "Episode 601: The Chocolate Curse," Planet Money, NPR, February 4, 2015, http://www.npr.org/sections/money/2015/02/04/383830776/episode-601-the-chocolate-curse.

90. Vicente Norero, "Bulletin from Balao," News Blog, Camino Verde, October 1, 2014, http://www.caminoverde.com.ec/bulletin-from-balao/.

COFFEE

1. "National Coffee Drinking Trends 2014," National Coffee Association USA, 13, accessed May 29, 2015, http://www.ncausa.org/i4a/pages/index.cfm?pageID=921.

2. "How Drugs Affect Neurotransmitters," The Brain from Top to Bottom, accessed March 23, 2014, http://thebrain.mcgill.ca/flash/i/i_03/i_03_m/i_03_m_par/i_03 _m_par_cafeine.html.

3. Marie Cheour, "Caffeine Effects on Different Parts of the Brain," Livestrong.com, August 16, 2013, http://www.livestrong.com/article/222993-caffeine-effects-on-different -parts-of-the-brain/.

4. "Ric Rhinehart: From Selling Better Coffee to Selling Coffee Better," YouTube video, 23:30, from the 2014 Specialty Coffee Association of America Symposium, posted by "SCAASymposium," May 13, 2014, https://www.youtube.com/watch?v=RZ rMxEztSrI.

5. "Central America—The Coffee Crisis: Effects and Strategies for Moving Forward," The World Bank, accessed February 19, 2014, http://go.worldbank.org/7N BUC645G0.

6. "Fairtrade and Coffee," Fairtrade Foundation, May 2012, 2, http://www.fairtrade .net/fileadmin/user_upload/content/2009/resources/2012_Fairtrade_and_coffee _Briefing.pdf.

7. "World Coffee Consumption," International Coffee Organization, last modified April 2015, http://ico.org/prices/new-consumption-table.pdf.

8. Lydia Zuraw, "Foodways: How Coffee Influenced the Course of History," *Morning Edition*, The Salt, NPR, April 24, 2013, http://www.npr.org/sections/thesalt /2013/04/24/178625554/how-coffee-influenced-the-course-of-history.

9. "Maillard Reactions," Food-Info, accessed February 19, 2014, http://www.food-info.net/uk/colour/maillard.htm.

10. Sarah Everts, "Processed: Food Science and the Modern Meal," *Chemical Heritage Magazine*, Fall 2013/Winter 2014, http://www.chemheritage.org/discover/media/magazine/articles/31-3-processed.aspx?page=3.

11. J. E. Hodge, "Dehydrated Foods, Chemistry of Browning Reactions in Model Systems," *Journal of Agricultural and Food Chemistry* 1, no. 15 (October 1953): 928–43, doi:10.1021/jf60015a004.

12. Joseph A. Rivera, "Under the Microscope: The Science of Browning Reactions," *Roast Magazine*, March/April 2008, 47, http://www.roastmagazine.com/resources/Articles/Roast_MarApr08_ScienceofBrowning.pdf.

13. Gordon J. Troup et al., "Stable Radical Content and Anti-Radical Activity of Roasted Arabica Coffee: From In-Tact Bean to Coffee Brew," *PLOS ONE* 10, no. 4 (April 2015): e0122834, 1–17, doi:10.1371/journal.pone.0122834; and Rivera, "Under the Microscope," 46.

14. Allison Aubrey, "Eating and Health: From Kale to Pale Ale, a Love of Bitter May Be in Your Genes," *Morning Edition*, The Salt, NPR, October 1, 2014, http://www.npr.org/sections/thesalt/2014/10/01/352771618/from-kale-to-pale-ale-a-love-of-bitter-may-be-in-your-genes.

15. Eliza Barclay, "The Gene for Sweet: Why We Don't All Taste Sugar the Same Way," *Morning Edition*, The Salt, NPR, July 24, 2015, http://www.npr.org/sections/thesalt/2015/07/24/425609156/the-gene-for-sweet-why-we-dont-all-taste-sugar-the-same-way.

16. A. L. Allen, J. E. McGeary, and J. E. Hayes, "Polymorphisms in TRPV1 and TAS2Rs Associate with Sensations from Sampled Ethanol," *Alcoholism: Clinical and Experimental Research* 38, no. 10 (October 2014): 2550–60, doi:10.1111/acer.12527.

17. Preston S. Wilson, "Coffee Roasting Acoustics," *Journal of the Acoustical Society of America* 135, no. 6 (June 2014): EL266, doi:10.1121/1.4874355.

18. Diane Ackerman, *A Natural History of the Senses* (New York: Vintage Books, 1995), 177.

19. NPR staff, "What Does Cold Sound Like? See If Your Ear Can Tell Temperature," *All Things Considered*, NPR, July 5, 2014, http://www.npr.org/2014/07/05/328842704/what-does-cold-sound-like.

20. Charles Spence and Maya U. Shankar, "The Influence of Auditory Cues on the Perception of, and Responses to, Food and Drink," *Journal of Sensory Studies* 25, no. 3 (March 2010): 419, doi:10.1111/j.1745-459X.2009.00267.x.

21. Charles Spence, "Eating with Our Ears: Assessing the Importance of the Sounds of Consumption on Our Perception and Enjoyment of Multisensory Flavour Experiences," *Flavour* 4, no. 3 (March 2015): 1–14, doi:10.1186/2044-7248-4-3.

22. Kimberly S. Yan and Robin Dando, "A Crossmodal Role for Audition in Taste Perception," *Journal of Experimental Psychology: Human Perception and Performance* 41, no. 3 (June 2015): 590–96, doi:10.1037/xhp0000044.

23. Nordic Nibbler, "Musical Tastes—Maaelstrøm," Food Studio, August 13, 2006, posted by Alisa Larsen, http://www.foodstudio.no/column/musical-tastes/.

24. Spence, "Eating with Our Ears."

25. Steph Ceraso, "Snap, Crackle, Pop: The Sonic Pleasures of Food," *Sounding Out!* (blog), July 14, 2014, http://soundstudiesblog.com/2014/07/14/snap-crackle-pop-the-sonic-pleasures-of-food/.

26. Spence and Shankar, "The Influence of Auditory Cues," 414.

27. "Coffee Tasting How-To's: Perfecting the Coffee Slurp," posted by BeanPoster, *The Pilot's Blog*, The Roasterie, January 10, 2013, http://www.theroasterie.com/blog/coffee-tasting-how-tos-perfecting-the-coffee-slurp/.

28. Mark Pendergrast, *Uncommon Grounds: The History of Coffee and How It Transformed Our World* (New York: Basic Books, 1999), 282.

29. Pendergrast, *Uncommon Grounds*, 281.

30. Sprudge staff, "Erna Knutsen Is a Specialty Coffee Living Legend. This Video Will Inspire You," Sprudge.com, May 15, 2014, http://sprudge.com/erna-knutsen-specialty-coffee-legend-video-will-inspire-56318.html.

31. "Baseline Data for the Conservation of Coffee Species," Kew Royal Botanical Gardens, accessed July 23, 2015, http://www.kew.org/science-conservation/research-data/science-directory/projects/baseline-data-conservation-coffee.

32. Julian Siddle and Vibeke Venema, "Saving Coffee from Extinction," *BBC News Magazine*, May 24, 2015, http://www.bbc.com/news/magazine-32736366.

33. "All You Need to Know About Coffee: Species and Varieties," CIRAD, accessed March 23, 2014, http://www.cirad.fr/en/publications-resources/science-for-all/the-issues/coffee/what-you-need-to-know/species-and-varieties.

34. Raimond Feil, "*Coffea*: Genus—Species—Varieties," trans. Kadri Koitsaar, genus-Coffea, September 20, 2010, https://genuscoffea.wordpress.com/coffea-article/; and Raf Aerts et al., "Genetic Variation and Risks of Introgression in the Wild *Coffea arabica* Gene Pool in South-Western Ethiopian Montane Rainforests," *Evolutionary Applications* 6, no. 2 (February 2013): 243, doi:10.1111/j.1752-4571.2012.00285.x.

35. "Protecting the Last Cloud Forests of Ethiopia," The Nature and Biodiversity Conservation Union (NABU), 2011, 2, https://www.nabu.de/downloads/international/eng/Projektblatt_IKI_%C3%84thiopien_eng_06122011.pdf; and "Part 1. Centers of Origins of Crop Plants and Agriculture," Bioversity International, accessed March 27, 2014, http://www.bioversityinternational.org/fileadmin/bioversity/publications/Web_version/47/ch06.htm.

36. Kenneth Davids, *Coffee: A Guide to Buying, Brewing, and Enjoying* (New York: St. Martin's Press, 2001), 12.

37. Stewart Lee Allen, *The Devil's Cup* (New York: Soho Press, 1999), 5.

38. *A Film About Coffee*, directed by Brandon Loper (San Francisco: Avocados and Coconuts, 2014), film, 6:25 (total running time 66 min.).

39. "Brazilian Coffee Beans," CoffeeResearch.org, Coffee Research Institute, accessed March 27, 2014, http://www.coffeeresearch.org/coffee/brazil.htm.

40. William Harrison Ukers, *All About Coffee* (New York: Tea and Coffee Trade Journal, 1922), 26.

41. Matthew Green, "The Lost World of the London Coffeehouse," *Public Domain Review*, accessed March 27, 2014, http://publicdomainreview.org/2013/08/07/the-lost-world-of-the-london-coffeehouse/#sthash.ypNOOlYK.dpuf.

42. Green, "The Lost World."

43. "The Tea Act," USHistory.org, Independence Hall Association, accessed March 28, 2014, http://www.ushistory.org/declaration/related/teaact.htm.

44. Andrew F. Smith, *The Oxford Encyclopedia of Food and Drink in America: A to J* (New York: Oxford University Press, 2004), 266.

45. "PBS—Black Coffee, Part 2 of 3—Gold in Your Cup," YouTube video, 57:44, from documentary televised by PBS, posted by "exotickd," April 6, 2012, https://www.youtube.com/watch?v=g9sp8Vj2anI.

46. History, "Coffee," *Modern Marvels episode guide*, season 12, aired September 14, 2005 (New York: A&E Television Networks, 2005), http://www.history.com/shows/modern-marvels/episodes/season-12.

47. "PBS—Black Coffee, Part 2 of 3."

48. Noah Zerbe, "Climate Change and Your Morning Joe," *Global Food Politics* (blog), September 12, 2013, https://globalfoodpolitics.wordpress.com/2013/09/12/climate-change-and-your-morning-joe/.

49. Metasebia Yoseph, *A Culture of Coffee* (CreateSpace Independent Publishing Platform, 2013), 13.

50. Aaron P. Davis et al., "The Impact of Climate Change on Indigenous Arabica Coffee (*Coffea Arabica*): Predicting Future Trends and Identifying Priorities," *PLOS ONE* 7, no. 11 (November 2012): e47981, 2, doi:10.1371/journal.pone.0047981.

51. Amanda Fiegl, "The Last Drop? Climate Change May Raise Coffee Prices, Lower Quality," *National Geographic News*, November 9, 2012, http://news.nationalgeographic.com/news/2012/11/121108-climate-change-coffee-coffea-arabica-botanical-garden-science/.

52. "Questions over Ethiopia's Coffee Crop," Ethiopia Forums, last modified October 5, 2011, http://ethiopiaforums.com/questions-over-ethiopias-coffee-crop/8680/.

53. Davis et al., "The Impact of Climate Change," 9.

54. Tim Schilling, "Why Genetic Diversity Matters," World Coffee Research, accessed July 24, 2015, http://worldcoffeeresearch.org/read-more/newsletter/161-why-genetic-diversity-matters.

55. "Aaron Davis: Arabica—From Origin to Extinction," YouTube video, 20:25, from a 2013 SCAA Symposium talk, posted by "SCAASymposium," May 6, 2013, https://www.youtube.com/watch?v=pr5Iis_J6-U.

56. Noah Zerbe, "Climate Change and Your Morning Joe," *Global Food Politics* (blog), September 12, 2013, https://globalfoodpolitics.wordpress.com/2013/09/12/climate-change-and-your-morning-joe/.

57. Austrian Academy of Sciences (OAW), United Nations Educational, Scientific and Cultural Organization (UNESCO), and Man and the Biosphere Programme (MAB), *Biosphere Reserves in the Mountains of the World: Excellence in the Clouds?* ed. Austrian MAB Committee (Vienna: Austrian Academy of Sciences Press, 2011), 89, http://www.unesco.org/new/fileadmin/MULTIMEDIA/HQ/SC/pdf/OAW_BR_Mountains_Excellence_in_the_Clouds_2011.pdf.

58. The World Bank, *Forests Sourcebook: Practical Guidance for Sustaining Forests in Development Cooperation* (Washington, DC: The International Bank for Reconstruction

and Development/The World Bank, 2008), 1, http://siteresources.worldbank.org/EXTFORSOUBOOK/Resources/completeforestsourcebookapril2008.pdf.

59. "Forest Loss in Kafa," The Nature and Biodiversity Conservation Union (NABU), accessed May 20, 2015, https://en.nabu.de/projects/ethiopia/kafa/area/17101.html.

60. Dave Goulson, A Sting in the Tale: My Adventures with Bumblebees (New York: Picador, 2013), 37.

61. David W. Roubik, "Tropical Agriculture: The Value of Bees to the Coffee Harvest," Nature 417 (June 2002): 708, doi:10.1038/417708a.

62. Nick Brown, "The Birds and the Bees: Coffee Quality and Yield in Tanzania," Daily Coffee News, Roast Magazine, February 11, 2014, http://dailycoffeenews.com/2014/02/11/the-birds-and-the-bees-coffee-quality-and-yield-in-tanzania/.

63. "FAQ's," International Bee Research Association (IBRA), accessed March 25, 2014, http://www.ibra.org.uk/categories/faq.

64. "Nearly One in 10 Wild Bee Species Face Extinction in Europe While the Status of More Than Half Remains Unknown—IUCN Report," International Union for Conservation of Nature (IUCN), March 19, 2015, http://www.iucn.org/?19073/Nearly-one-in-ten-wild-bee-species-face-extinction-in-Europe-while-the-status-of-more-than-half-remains-unknown-IUCN-report.

65. Kim Kaplan, "Bee Survey: Lower Winter Losses, Higher Summer, Increased Total Annual Losses," USDA Agricultural Research Service, May 13, 2015, http://ars.usda.gov/is/pr/2015/150513.htm.

66. Ken Thompson, "The Nerve Poison Harming Our Bees," The Telegraph, May 11, 2012, http://www.telegraph.co.uk/gardening/9257431/The-nerve-poison-harming-our-bees.html.

67. Maj Rundlöf et al., "Seed Coating with a Neonicotinoid Insecticide Negatively Affects Wild Bees," Nature 521 (May 2015): 77–80, doi:10.1038/nature14420.

68. Thompson, "The Nerve Poison Harming Our Bees."

69. "Peter Raven on Biodiversity: Science and Its Context," The Biodiversity Leadership Awards, accessed March 25, 2014, http://www.biodiversityleadershipawards.org/raven.htm.

70. Lars Hein and Franz Gatzweiler, "The Economic Value of Coffee (Coffea arabica) Genetic Resources," Ecological Economics 60 (November 2006): 183, doi:10.1016/j.ecolecon.2005.11.022.

71. Julian Siddle and Vibeke Venema, "Saving Coffee from Extinction," BBC News Magazine, May 24, 2015, http://www.bbc.com/news/magazine-32736366.

72. Bill McKibben, "Welcome to the Climate Crisis," Washington Post, May 27, 2006, http://www.washingtonpost.com/wp-dyn/content/article/2006/05/26/AR2006052601549.html.

73. James McWilliams, "Coffee Leaf Rust: It's Coming for Your Morning Joe," The Atlantic, May 21, 2013, http://www.theatlantic.com/international/archive/2013/05/coffee-leaf-rust-its-coming-for-your-morning-joe/276084/.

74. Jennifer Bingham Hull, "Can Coffee Drinkers Save the Rain Forest?" The Atlantic, August 1999, http://www.theatlantic.com/magazine/archive/1999/08/can-coffee-drinkers-save-the-rain-forest/377733/.

75. Hull, "Can Coffee Drinkers Save the Rain Forest?"

76. McWilliams, "Coffee Leaf Rust."

77. "Coffee from Around the World," National Coffee Association USA, accessed March 25, 2014, http://ncausa.org/i4a/pages/index.cfm?pageid=75.

78. Julie Craves, "How 'Wild' Is Ethiopian Forest Coffee?" Coffee & Conservation, February 6, 2011, last modified March 23, 2014, http://www.coffeehabitat.com/2011 /02/ethiopia-wild-forest-coffee/.

79. Christine B. Schmitt et al., "Wild Coffee Management and Plant Diversity in the Montane Rainforest of Southwestern Ethiopia," *African Journal of Ecology* 48, no. 1 (March 2010): 78–86, doi:10.1111/j.1365-2028.2009.01084.x.

80. Craves, "How 'Wild' Is Ethiopian Forest Coffee?"

81. Craves, "How 'Wild' Is Ethiopian Forest Coffee?"

82. Craves, "How 'Wild' Is Ethiopian Forest Coffee?"

83. Paul Hicks, "We All Drink Downstream," *Coffeelands* (blog), Catholic Relief Services, April 17, 2015, http://coffeelands.crs.org/2015/04/we-all-drink-downstream/.

84. Adam Smith, *An Inquiry into the Nature and Causes of the Wealth of Nations*, ed. Kathryn Sutherland (New York: Oxford University Press, 1993), 14.

85. Christopher M. Bacon et al., eds., *Confronting the Coffee Crisis: Fair Trade, Sustainable Livelihoods and Ecosystems in Mexico and Central America* (Cambridge, MA: MIT Press, 2008), 44.

86. Jared Cummans, "The Ultimate Guide to Coffee Investing," Commodity HQ, June 7, 2011, http://commodityhq.com/2011/the-ultimate-guide-to-coffee-investing/.

87. Charis Gresser and Sophia Tickell, *Mugged: Poverty in Your Cup* (Boston, MA: Oxfam Publishing, 2004), 17, http://www.oxfamamerica.org/explore/research-publications /mugged-poverty-in-your-coffee-cup/.

88. Gresser and Tickell, *Mugged*, 17.

89. "History," International Coffee Organization, accessed March 31, 2014, http://www .ico.org/icohistory_e.asp.

90. *A Film About Coffee*, directed by Brandon Loper (San Francisco: Avocados and Coconuts, 2014), film, 6:25 (total running time 66 min.).

91. "Black Gold: The Economics of Coffee," *Independent Lens*, PBS, accessed March 31, 2014, http://www.pbs.org/independentlens/blackgold/economics.html.

92. Dow Jones, "Coffee Perks Up Commodity-Futures Markets," *Australian Business Review*, January 1, 2015, http://www.theaustralian.com.au/business/news/coffee-perks-up-commodity-futures-markets/story-e6frg906-1227171768849.

93. "Fairtrade and Coffee," Fairtrade Foundation, May 2012, 10, http://www.fairtrade .net/fileadmin/user_upload/content/2009/resources/2012_Fairtrade_and_coffee _Briefing.pdf.

94. "Black Gold: The Economics of Coffee."

95. "Fairtrade and Coffee," 7.

96. Tim Schilling, "Why Genetic Diversity Matters," World Coffee Research, accessed July 24, 2015, http://worldcoffeeresearch.org/read-more/newsletter/161-why-genetic-diversity-matters.

97. Eileen P. Kenny, "Aaron Wood," Birds of Unusual Vitality, accessed June 21, 2015, http://cargocollective.com/birdsofuv/Aaron-Wood.

98. Silvio Leite, AgriCafé, *Cup of Excellence Cupping Calibration*, online video, 48:13, posted by "NBC Press," September 5, 2013 (Copenhagen: Nordic Barista Cup videos, 2013), http://nordicbaristacup.com/2013/09/cup-of-excellence-cupping-calibration-by-silvio-leite-agricafe/.

99. "Cupping Protocols," Resources, Specialty Coffee Association of America, accessed March 23, 2014, http://www.scaa.org/?page=resources&d=cupping-protocols.

100. Sarah Allen, "Counter Culture's Tim Hill Sheds Light on the New Taster's Flavor Wheel," *Barista Magazine Blog*, January 6, 2014, http://baristamagazine.com/blog/counter-cultures-tim-hill-sheds-light-on-the-new-tasters-flavor-wheel/.

BEER

1. Garrett Oliver, "Brewer Toasts," in Rob Hill, *Total Wine & More: Total Guide to Beer* (Potomac, MD: Retail Services & Systems, 2012), 183.

2. *Encyclopedia Britannica Online*, s.v. "Cannabaceae," accessed June 27, 2015, http://www.britannica.com/plant/Cannabaceae.

3. Randy Mosher, *Tasting Beer: An Insider's Guide to the World's Greatest Drink* (North Adams, MA: Storey Publishing, 2009), 1.

4. Lydia Saad, "Beer Is Americans' Adult Beverage of Choice This Year," Gallup, July 23, 2014, http://www.gallup.com/poll/174074/beer-americans-adult-beverage-choice-year.aspx.

5. Carl Sagan, *Cosmos* (New York: Random House, 1980), 233.

6. Andrew Bennett Hellman, "Gut Bacteria Gene Complement Dwarfs Human Genome," *Nature*, March 3, 2010, http://www.nature.com/news/2010/100303/full/news.2010.104.html; and Richard Conniff, "Microbes: The Trillions of Creatures Governing Your Health," *Smithsonian Magazine*, May 2013, http://www.smithsonianmag.com/science-nature/microbes-the-trillions-of-creatures-governing-your-health-37413457/?all.

7. S. E. Gould, "The Bacteria in Breast Milk," Blogs, *Scientific American*, December 8, 2013, http://blogs.scientificamerican.com/lab-rat/the-bacteria-in-breast-milk/.

8. Ann Reid and Michael Ingerson-Mahar, *If the Yeast Ain't Happy, Ain't Nobody Happy*, a report from the American Academy of Microbiology (Washington, DC: American Academy of Microbiology, 2012), http://csibrewers.org/content/wp-content/uploads/2013/03/MicrobiologyOfBeer.pdf.

9. "The Benefits of Yeast," Explore Yeast, Lessaffre, accessed December 5, 2014, http://www.exploreyeast.com/benefits-yeast.

10. Rowan Jacobsen, "The Most Delicious Lab: How a Pair of Harvard Microbiologists Became Accidental Cheese Celebrities," *Boston Globe*, October 23, 2011, http://www.bostonglobe.com/ideas/2011/10/22/survivor-piece-cheese/36e0KwiPYHAw9H6HoqBzNP/story.html.

11. Mosher, *Tasting Beer*, 54.

12. Fred Shapiro, "Quotes Uncovered: Beer or Wine as Proof?" *Freakonomics* (blog), March 24, 2011, http://freakonomics.com/2011/03/24/quotes-uncovered-beer-or-wine-as-proof/.

13. Charles Bamforth, *Beer: Tap into the Art and Science of Brewing* (New York: Oxford University Press, 2003), 141.

14. Ian Spencer Hornsey, *A History of Beer and Brewing* (Cambridge: The Royal Society of Chemistry: 2003), 601.

15. Hornsey, *A History of Beer and Brewing,* 601.

16. Richard Twyman, "Model Organisms: Yeast," The Human Genome, Wellcome Trust, August 28, 2002, http://genome.wellcome.ac.uk/doc_WTD020808.html.

17. S. Michal Jazwinski, "Yeast," *Medicine Encyclopedia,* accessed December 5, 2014, http://medicine.jrank.org/pages/1863/Yeast.html.

18. Jacobsen, "The Most Delicious Lab."

19. Camilo Mora et al., "How Many Species Are There on Earth and in the Ocean?" *PLOS Biology* 9, no. 8 (August 2011): e1001127, 1–8, doi:10.1371/journal.pbio.1001127.

20. "The Extinction Crisis," Center for Biological Diversity, accessed December 10, 2014, http://www.biologicaldiversity.org/programs/biodiversity/elements_of_biodiversity/extinction_crisis/.

21. The World Bank, *Forests Sourcebook: Practical Guidance for Sustaining Forests in Development Cooperation* (Washington, DC: The International Bank for Reconstruction and Development/The World Bank, 2008), 1, http://siteresources.worldbank.org/EXTFORSOUBOOK/Resources/completeforestsourcebookapril2008.pdf; and "Chimpanzees: Biology & Habitat," The Jane Goodall Institute, accessed June 27, 2015, http://www.janegoodall.org/chimpanzees/biology-habitat.

22. "Extinction Threat Growing for Mankind's Closest Relatives," news release, International Union for Conservation of Nature (IUCN), August 3, 2008, http://www.iucn.org/?1391/1/Extinction-threat-growing-for-mankinds-closest-relatives.

23. Cell Press, "Thanks, Fruit Flies, for That Pleasing Beer Scent," news release, The American Association for the Advancement of Science, October 9, 2014, http://www.eurekalert.org/pub_releases/2014-10/cp-tff100214.php.

24. Irene Stefanini et al., "Role of Social Wasps in *Saccharomyces cerevisiae* Ecology and Evolution," *Proceedings of the National Academy of Sciences of the United States of America* (PNAS) 109, no. 33 (August 2012): 13398–403, doi:10.1073/pnas.1208362109.

25. Frank DiGennaro, "The Coolships Have Landed," *Table Matters,* February 14, 2014, http://tablematters.com/2014/02/14/the-coolships-have-landed/.

26. Freek Spitaels et al., "The Microbial Diversity of Traditional Spontaneously Fermented Lambic Beer," *PLOS Biology* 9, no. 4 (April 2014): e95384, 1–13, doi:10.1371/journal.pone.0095384.

27. Gregg Glaser, "In Search of Lambic: The World's Oldest Beer Style," *All About Beer Magazine,* July 1, 2001, http://allaboutbeer.com/article/in-search-of-lambic/.

28. Martyn Cornell, "When Brick Lane Was Home to the Biggest Brewery in the World," *Zythophile* (blog), March 14, 2013, http://zythophile.co.uk/2013/03/14/when-brick-lane-was-home-to-the-biggest-brewery-in-the-world/.

29. Cornell, "When Brick Lane Was Home."

30. Cornell, "When Brick Lane Was Home."

31. Cornell, "When Brick Lane Was Home."

32. "Established 1666, Re-Established 2010," About, Truman's, accessed January 20, 2015, http://www.trumansbeer.co.uk/about/.

33. "Established 1666, Re-Established 2010."

34. Mosher, *Tasting Beer*, 45.

35. Mosher, *Tasting Beer*, 39.

36. Ralph Riley, "Obituary: Douglas Bell," *The Independent*, July 17, 1993, http://www.independent.co.uk/news/people/obituary-douglas-bell-1485432.html.

37. Richard Mithen, "East Anglia: Plant Breeders Celebrate Legacy of the Plant Breeding Institute," *East Anglian Daily Times*, June 26, 2012, http://www.eadt.co.uk/news/east_anglia_plant_breeders_celebrate_legacy_of_the_plant_breeding_institute_1_1420514.

38. "Maris Otter Malt," Brew-Engine, accessed January 20, 2015, http://brew-engine.com/ingredients/grains_and_sugars/maris_otter_malt.html.

39. John Palmer, "Chapter 12—What Is Malted Grain?," *How to Brew*, 2009, accessed January 20, 2015, http://www.howtobrew.com/section2/chapter12.html.

40. "Malting Barley Update: Summer 2014," The Maltsters' Association of Great Britain (MAGB), accessed January 20, 2015, http://www.ukmalt.com/sites/default/files/files/MAGB%20MBUpdate14%20vL.pdf.

41. William Bostwick, *The Brewer's Tale: A History of the World According to Beer* (New York: W. W. Norton, 2014), 147.

42. Garrett Oliver, *The Oxford Companion to Beer* (New York: Oxford University Press, 2012), 571.

43. "Maris Otter Malt Extract Syrup—Made with Crisp Maris Otter," More Beer!, accessed January 20, 2015, http://www.morebeer.com/products/maris-otter-malt-extract-syrup-crisp.html.

44. "John Innes Centenary Beer," John Innes Centre, accessed January 14, 2015, https://www.jic.ac.uk/centenary/beer.htm.

45. "NCGR-Corvallis—*Humulus* Germplasm," USDA Agricultural Research Service, last modified December 28, 2014, https://www.ars.usda.gov/News/docs.htm?docid=11069.

46. Horst Dornbusch, "A Short History of *Humulus lupulus*," *Brew Your Own*, January/February 2015, https://byo.com/index.php?option=com_mijoshop&Itemid=596&route=product/product&manufacturer_id=1&product_id=181&page=1&limit=25.

47. William Bostwick, "The Bitter Truth," *Aeon Magazine*, November 13, 2014, http://aeon.co/magazine/culture/why-are-todays-craft-beers-so-bitter/?utm_source=Aeon+newsletter&utm_campaign=a04ca3e2da-Daily_newsletter_13_November_201411_13_2014&utm_medium=email&utm_term=0_411a82e59d-a04ca3e2da-59906629.

48. Martyn Cornell, "How Long Have English Brewers Been Using American Hops? Much Longer Than You Think," *Zythophile* (blog), May 27, 2013, http://zythophile.co.uk/2013/05/27/how-long-have-english-brewers-been-using-american-hops-much-longer-than-you-think/.

49. "History of Craft Brewing," Brewers Association, accessed January 24, 2015, https://www.brewersassociation.org/brewers-association/history/history-of-craft-brewing/.

50. "Beer Slogans," Brookston Beer Bulletin, accessed January 24, 2015, http://brookstonbeerbulletin.com/beer-slogans/.

51. "History of Craft Brewing."

52. "History of Beer: The Revival," Craft Beer, Brewers Association, accessed June 15, 2015, http://www.craftbeer.com/the-beverage/history-of-beer/the-revival.

53. Mosher, *Tasting Beer*, 23.

54. Mosher, *Tasting Beer*, 3.

55. Natasha Geiling, "In Search of the Great American Beer," Smithsonian.com, July 30, 2014, http://www.smithsonianmag.com/arts-culture/search-great-american-beer-180951966/?no-ist.

56. Geiling, "In Search of the Great American Beer."

57. Brad Smith, "Noble Hops for European Beer Styles," *BeerSmith Home Brewing Blog*, February 5, 2012, http://beersmith.com/blog/2012/02/05/noble-hops-for-european-beer-styles/.

58. "Bitterness," *Beer Sensory Science* (blog), November 6, 2010, https://beersensoryscience.wordpress.com/2010/11/06/bitterness/.

59. Marc Bona, "Beer 101: India Pale Ale's Origin Is Rooted in Geography, History," Cleveland.com, October 15, 2013, http://www.cleveland.com/taste/index.ssf/2013/10/india_pale_ales_origin_is_root.html.

60. Maggie Dutton, "What's the Difference Between Pale Ale and IPA?" *Seattle Weekly*, August 19, 2008, http://www.seattleweekly.com/2008-08-20/food/pale-ale-vs-ipa/.

61. Geiling, "In Search of the Great American Beer."

62. J. Robert Sirrine et al., *Sustainable Hop Production in the Great Lakes Region*, extension bulletin E-3083 (East Lansing: Michigan State University Extension, January 2010), http://www.uvm.edu/extension/cropsoil/wp-content/uploads/Sirrine-Sustainable-Hop-Production-in-the-Great-Lakes-Region.pdf.

63. David H. Gent et al., eds., *Field Guide for Integrated Pest Management in Hops*, 2nd ed. (Moxee, WA: Washington Hop Commission, 2010), http://www.ars.usda.gov/SP2UserFiles/person/37109/HopHandbook2010.pdf.

64. C. B. Skotland and Dennis A. Johnson, "Control of Downy Mildew of Hops," *Plant Disease* 67, no. 11 (November 1983): 1183–85, http://www.apsnet.org/publications/plantdisease/backissues/Documents/1983Articles/PlantDisease67n11_1183.pdf.

65. "*Humulus lupulus* var. *neomexicanus*," Integrated Taxonomic Information System online database, accessed December 15, 2014, http://www.itis.gov/servlet/SingleRpt/SingleRpt?search_topic=TSN&search_value=528499.

66. "About," El Dorado Hops, CLS Farms, accessed December 15, 2014, http://eldoradohops.com/about.html.

67. "Hop Growing in Washington," USAHops.org, Hop Growers of America, accessed December 15, 2014, http://www.usahops.org/index.cfm?fuseaction=hop_farming&pageID=13.

68. "Brewers Association Lists Top 50 Breweries of 2014," Brewers Association, accessed

June 15, 2015, https://www.brewersassociation.org/press-releases/brewers-association-lists-top-50-breweries-of-2014/.

69. "Harvest Wild Hop IPA—Neomexicanus Varietal," Sierra Nevada, accessed December 20, 2014, http://www.sierranevada.com/beer/specialty/harvest-wild-hop-ipa-neomexicanus-varietal.

70. "Craft Brewer Defined," Brewers Association, accessed January 24, 2015, https://www.brewersassociation.org/statistics/craft-brewer-defined/.

71. "The Rise of Craft Beer in the US—Craft Beer Sales Have Doubled in the Past Six Years and Are Set to Triple by 2017," Mintel, January 23, 2013, http://www.mintel.com/press-centre/food-and-drink/the-rise-of-craft-beer-in-the-us-craft-beer-sales-have-doubled-in-the-past-six-years-and-are-set-to-triple-by-2017.

72. Tripp Mickle, "Bud Crowded Out by Craft Beer Craze," *Wall Street Journal*, November 23, 2014, http://www.wsj.com/articles/budweiser-ditches-the-clydesdales-for-jay-z-1416784086.

73. "The Rise of Craft Beer in the US."

74. Jon Terbush, "Is Craft Beer Killing Bud Light?" *The Week*, July 8, 2014, http://theweek.com/speedreads/450344/craft-beer-killing-bud-light.

75. Mosher, *Tasting Beer*, 1.

76. Steve Hindy, *The Craft Beer Revolution: How a Band of Microbreweries Is Transforming the World's Favorite Drink* (New York: Palgrave Macmillan Trade, 2014), 128.

77. "Reinventing the Wheel," *Brewers' Guardian*, March/April 2010, http://www.flavoractiv.com/wpcms/wp-content/uploads/2010/11/Brewers-Guardian-Article-March-April-2010-Reinventing-the-Wheel.pdf.

78. Mosher, *Tasting Beer*, 53.

79. Mosher, *Tasting Beer*, 56.

80. Ann Reid, Charles Bamforth, and Rebecca Newman, "The Microbiology of Beer" ("Microbes After Hours" series video presentation sponsored by the American Society for Microbiology, Washington, DC, October 10, 2013).

81. Nazlin Imram, "The Role of Visual Cues in Consumer Perception and Acceptance of a Food Product," *Nutrition & Food Science* 99, no. 5 (October 1999): 226, doi:10.1108/00346659910277650.

82. Susan A. Langstaff and M. J. Lewis, "Foam and the Perception of Beer Flavor and Mouthfeel," MBAA *Technical Quarterly* 30, no. 1 (1993): 16–17, http://appliedsensory.com/uploads/Foam.pdf.

83. Mosher, *Tasting Beer*, 64.

84. Jay A. Gottfried and Raymond J. Dolan, "The Nose Smells What the Eye Sees: Cross-modal Visual Facilitation of Human Olfactory Perception," *Neuron* 39, no. 2 (July 2003): 375–86, doi:10.1016/S0896-6273(03)00392-1.

85. "Aldi Confirms up to 100% Horsemeat in Beef Products," *The Guardian*, February 9, 2013, http://www.theguardian.com/business/2013/feb/09/aldi-100-percent-horsemeat-beef-products.

86. Diane Ackerman, *A Natural History of the Senses* (New York: Vintage Books, 1995), 230.

87. Harvey Lodish et al., "Section 22.3: Collagen: The Fibrous Proteins of the Matrix" in *Molecular Cell Biology*, 4th ed. (New York: W. H. Freeman, 2000).

88. Ackerman, *A Natural History of the Senses*, 254.

89. Brad Smith, "Beer Color: Understanding SRM, Lovibond and EBC," *BeerSmith Home Brewing Blog*, April 29, 2008, http://beersmith.com/blog/2008/04/29/beer-color-understanding-srm-lovibond-and-ebc/.

90. Ray Daniels, "Beer Color Demystified," More Beer!, July 15, 2012, http://www.more beer.com/articles/beercolor.

91. JoAndrea Hoegg and Joseph W. Alba, "Taste Perception: More Than Meets the Tongue," *Journal of Consumer Research* 33, no. 4 (March 2007): 493, doi:10.1086/510222.

92. Massimiliano Zampini et al., "The Multisensory Perception of Flavor: Assessing the Influence of Color Cues on Flavor Discrimination Responses," *Food Quality and Preference* 18, no. 7 (October 2007): 976–77, doi:10.1016/j.foodqual.2007.04.001.

93. Zampini et al., "The Multisensory Perception of Flavor," 976.

94. Zampini et al., "The Multisensory Perception of Flavor," 981.

95. Langstaff and Lewis, "Foam and the Perception of Beer Flavor and Mouthfeel."

96. Mosher, *Tasting Beer*, 105.

97. G. Donadini, M. D. Fumi, and M. D. de Faveri, "How Foam Appearance Influences the Italian Consumer's Beer Perception and Preference," *Journal of the Institute of Brewing* 117, no. 4 (January 2011): 528, doi:10.1002/j.2050-0416.2011.tb00500.x.

98. J. E. Smythe, M. A. O'Mahony, and C. W. Bamforth, "The Impact of the Appearance of Beer on Its Perception," *Journal of the Institute of Brewing* 108, no. 1 (January 2002): 39, doi:10.1002/j.2050-0416.2002.tb00120.x.

99. Smythe, O'Mahony, and Bamforth, "The Impact of the Appearance of Beer," 41.

100. L. Blasco, M. Viñas, and T. G. Villa, "Proteins Influencing Foam Formation in Wine and Beer: The Role of Yeast," *Internal Microbiology* 14, no. 2 (June 2011): 61–71, http://www.ncbi.nlm.nih.gov/pubmed/22069150.

101. Emma Christensen, "Good Foam, Bad Foam: What's the Deal with Beer Foam?" *The Kitchn*, September 27, 2011, http://www.thekitchn.com/good-foam-bad-foam-whats-the-d-156966.

102. Stan Hieronymus, "The Dirt on Terroir," *Draft*, July 26, 2011, http://draftmag.com/the-dirt-on-terroir/.

103. Robert J. Braidwood et al., "Symposium: Did Man Once Live by Beer Alone?" *American Anthropologist* 55, no. 4 (October 1953): 516, http://www.jstor.org/stable/663781.

104. Jeffrey P. Kahn, "How Beer Gave Us Civilization," Grey Matter, *New York Times*, March 15, 2013, http://www.nytimes.com/2013/03/17/opinion/sunday/how-beer-gave-us-civilization.html?_r=0.

105. Solomon H. Katz and Mary M. Voigt, "Bread and Beer: The Early Use of Cereals in the Human Diet," *Expedition* 28, no. 2 (1986): 27.

106. Abigail Tucker, "The Beer Archaeologist," *Smithsonian Magazine*, August 2011, http://www.smithsonianmag.com/history/the-beer-archaeologist-17016372/?no-ist.

107. Paul T. Nicholson and Ian Shaw, eds., *Ancient Egyptian Materials and Technology* (Cambridge: University of Cambridge Press, 2000), 570.

108. "Ninkasi, the Sumerian Goddess of Brewing and Beer," *Beer Advocate*, December 20, 2000, accessed December 20, 2014, http://www.beeradvocate.com/articles/304/.

109. Mosher, *Tasting Beer*, 7.

110. Niek Veldhuis, email message to author, May 7, 2015; and Andrew R. George, *The Babylonian Gilgamesh Epic: Introduction, Critical Edition and Cuneiform Texts* (Oxford: Oxford University Press, 2003), 148–49.

111. Paige Latham, "Women in Beer: From Sumeria to Summit," Alcohol by Volume, July 28, 2015, http://www.alcoholbyvolumeblog.com/?p=8435.

112. Jenny Hill, "Tjenenet," Ancient Egypt Online, accessed December 20, 2014, http://www.ancientegyptonline.co.uk/tjenenet.html; and Joshua J. Mark, "Beer," *Ancient History Encyclopedia*, March 1, 2011, http://www.ancient.eu/Beer/.

113. Mosher, *Tasting Beer*, 8.

114. Rob Hill, *Total Wine & More: Total Guide to Beer* (Potomac, MD: Retail Services & Systems, 2012), 11.

115. "All Hail Ninkasi!" *Pink Boots Society*, January 25, 2013, http://pinkbootssociety.org/2013/01/25/all-hail-ninkasi/.

116. Mosher, *Tasting Beer*, 146.

117. Hill, *Total Wine & More*, 13.

118. Judith M. Bennett, *Ale, Beer, and Brewsters in England: Women's Work in a Changing World* (Oxford: Oxford University Press, 1996), 15.

119. Iain Gately, *Drink: A Cultural History of Alcohol* (New York: Gotham Books, 2008), 81.

120. Brian Glover, *The Lost Beers & Breweries of Britain* (Stroud, Gloucestershire: Amberley Publishing, 2013), in the chapter "Joule's Stone Ale," electronic edition.

121. Martin H. Stack, "A Concise History of America's Brewing Industry," EH.net, Economic History Association, July 4, 2003, http://eh.net/encyclopedia/a-concise-history-of-americas-brewing-industry/.

122. "Early History," Women's Christian Temperance Union, accessed January 7, 2015, http://www.wctu.org/earlyhistory.html.

123. Kenneth D. Rose, *American Women and the Repeal of Prohibition* (New York: New York University Press, 1996).

124. Carl Miller, "Beer and Television: Perfectly Tuned In," *All About Beer Magazine*, January 1, 2002, http://allaboutbeer.com/article/beer-and-television/.

125. "'Gotta Grab a Heini' Heineken Commercial," YouTube video, 00:30, televised Heineken commercial, posted by "Eric Orman," March 26, 2012, https://www.youtube.com/watch?v=klXegSorqHM.

126. Saad, "Beer Is Americans' Adult Beverage of Choice."

127. Mosher, *Tasting Beer*, 97.

128. Mosher, *Tasting Beer*, 105.

129. Mosher, *Tasting Beer*, 95.

130. Jeff Leach, "American Gut Releases Latest Results and Pipeline," American Gut, July 9, 2014, http://americangut.org/?p=189.

BREAD

1. Vishwa Mohan, "Punjab Continues to Be the Wheat Bowl of India," *Times of India*, May 21, 2014, http://timesofindia.indiatimes.com/india/Rabi-marketing-Season-foodgrain-production/articleshow/35437468.cms?.

2. "Introduction to Punjab," Department of Agriculture, Government of Punjab, accessed April 7, 2015, http://agripb.gov.in/home.php?page=intpu.

3. Food and Agriculture Organization of the United Nations (FAO), "Staple Foods: What Do People Eat?" Corporate Document Repository, accessed April 7, 2015, http://www.fao.org/docrep/u8480e/u8480e07.htm.

4. FAO, "Staple Foods: What Do People Eat?"

5. "*Triticum aestivum* (Bread Wheat)," Kew Royal Botanic Gardens, accessed April 7, 2015, http://www.kew.org/science-conservation/plants-fungi/triticum-aestivum-bread-wheat.

6. Stan Cox, "Types of Wheat: What to Grow and How to Use It," *Mother Earth News*, February/March 2014, http://www.motherearthnews.com/real-food/types-of-wheat-zm0z14fmzmat.aspx?PageId=1.

7. Cox, "Types of Wheat."

8. "The Biology of *Triticum aestivum* L. (Wheat)," Canadian Food Inspection Agency, last modified March 5, 2012, http://www.inspection.gc.ca/plants/plants-with-novel-traits/applicants/directive-94-08/biology-documents/triticum-aestivum-l-/eng/1330 982915526/1330982992440#B4.

9. "The Biology of *Triticum turgidum* ssp. *durum* (Durum Wheat)," Canadian Food Inspection Agency, last modified July 2, 2014, http://www.inspection.gc.ca/plants/plants-with-novel-traits/applicants/directive-94-08/biology-documents/triticum-turgidum-ssp-durum/eng/1330983955477/1330984025320#B4.

10. Cox, "Types of Wheat"; and "*Triticum aestivum* (Bread Wheat)."

11. "*Triticum aestivum* (Bread Wheat)."

12. "Wheat's Role in the U.S. Diet," USDA Economic Research Service, last modified June 19, 2013, http://www.ers.usda.gov/topics/crops/wheat/wheats-role-in-the-us-diet.aspx.

13. Kamala Elizabeth Nayar and Jaswinder Singh Sandhu, *The Socially Involved Renunciate: Guru Nanak's Discourse to the Nath Yogis* (Albany: State University of New York Press, 2007), 112.

14. "Mool Mantar," Guru Gobind Singh Gurdwara, accessed April 9, 2015, http://www.bradfordgurdwara.com/intro-to-sikhism/mool-mantar/.

15. Naazneen Karmali, "For the First Time, India's 100 Richest of 2014 Are All Billionaires," *Forbes*, September 24, 2014, http://www.forbes.com/sites/naazneenkarmali/2014/09/24/indias-100-richest-of-2014-are-all-billionaires-for-the-first-time/; and Nilanjana Bhowmick, "India Is Home to More Poor People than Anywhere Else on Earth," *Time*, July 17, 2014, http://time.com/2999550/india-home-to-most-poor-people/.

16. Cameron Stauch, "Golden Temple Sikhs Put Their Faith in Feeding 100,000 a Day," *Zester Daily*, July 12, 2013, http://zesterdaily.com/world/golden-temple-sikhs-put-faith-in-community-kitchens/.

17. Diane Ackerman, *A Natural History of the Senses* (New York: Random House, 1995), 128.

18. Maria Konnikova, "The Power of Touch," *New Yorker*, March 4, 2015, http://www.newyorker.com/science/maria-konnikova/power-touch.

19. "Fingertips to Hair Follicles: Why 'Touch' Triggers Pleasure and Pain," *Fresh Air*, NPR, February 3, 2015, http://www.npr.org/sections/health-shots/2015/02/03/383426166/fingertips-to-hair-follicles-why-touch-causes-pleasure-and-pain.

20. "Fingertips to Hair Follicles."

21. Monica Driscoll and Joshua Kaplan, "Touch Receptor Development and Differentiation," chap. 23, sec. 3, in *C. elegans* II, 2nd ed., ed. Donald L. Riddle et al. (Cold Spring Harbor, NY: Cold Spring Harbor Laboratory Press, 1997).

22. Dale Purves et al., eds., "Mechanoreceptors Specialized to Receive Tactile Information," chap. 9, in *Neuroscience*, 2nd ed. (Sunderland, MA: Sinauer Associates, 2001); and Ian Darian-Smith and Kenneth O. Johnson, "Thermal Sensibility and Thermoreceptors," *Journal of Investigative Dermatology* 69, no. 1 (July 1977): 146–53, doi:10.1111/1523-1747.ep12497936.

23. Elise Hancock, "A Primer on Touch," *Johns Hopkins Magazine*, September 1996, http://pages.jh.edu/~jhumag/996web/touch.html.

24. Kelsey Nowakowski, "International Women's Day 2014: Revealing the Gap Between Men and Women Farmers," *National Geographic*, March 7, 2014, http://news.nation algeographic.com/news/2014/03/140308-international-female-farmers/.

25. Alice Peck, ed., *Bread, Body, Spirit: Finding the Sacred in Food* (Woodstock, VT: SkyLight Paths, 2008), 68.

26. Sukhmandir Khalsa, "Kirpan Defined: The Sikh Ceremonial Short Sword," About Religion, About.com, accessed April 9, 2015, http://sikhism.about.com/od/glossary /g/Kirpan.htm.

27. "FAQ: Equality," Real Sikhism, accessed April 9, 2015, http://realsikhism.com/index .php?subaction=showfull&id=1248311895&ucat=7.

28. Food and Agriculture Organization of the United Nations (FAO), "Milk Facts," last modified January 7, 2015, http://www.fao.org/resources/infographics/infographics-details/en/c/273893/?utm_source=twitter&utm_medium=social+media&utm _campaign=faoknowledge.

29. John Barry and E. Gene Frankland, eds., *International Encyclopedia of Environmental Politics* (New York: Routledge, 2002), 393.

30. Rekha Sharma et al., "Genetic Diversity and Relationship of Cattle Populations in East India: Distinguishing Lesser Known Cattle Populations and Established Breeds Based on STR Markers," *Springerplus* 2 (July 2013): 359, doi:10.1186/2193-1801-2-359.

31. Devinder Sharma, "Holy Cows—Acclaimed Abroad, Despised at Home," *Ground Reality* (blog), September 6, 2010, http://devinder-sharma.blogspot.in/2010/09/holy-cows-acclaimed-abroad-despised-at.html.

32. Sharma, "Holy Cows."

33. Shashi Tharoor, "The Ugly Briton," *Time*, November 29, 2010, http://content.time .com/time/magazine/article/0,9171,2031992,00.html.

34. International Food Policy Research Institute (IFPRI), *Green Revolution: Curse or Blessing?*, issue brief (Washington, DC: IFPRI, 2002), 2, http://ebrary.ifpri.org/cdm/ref /collection/p15738coll2/id/64639.

35. "Remarks by Administrator Rajiv Shah to the CGIAR Board of Directors," USAID, December 7, 2012, http://www.usaid.gov/news-information/speeches/remarks-admin istrator-rajiv-shah-cgiar-board-directors.

36. Prabir Kumar De, ed., *Public Policy and Systems* (New Delhi: Dorling Kindersley, 2012), 72.

37. IFPRI, *Green Revolution: Curse or Blessing?*, 3.

38. Mohammad Ali Alaeddini and Majid Olia, "Green Revolution in India; An Experience," in *Proceedings of the Fourth International Iran & Russia Conference* (Shahrekord, Iran: International Iran and Russia Conference, 2004), 1119, 1121.

39. M. S. Swaminathan, "How to End Hunger: Lessons from the Father of India's Green Revolution," *The Guardian*, April 30, 2015, http://www.theguardian.com/global-development-professionals-network/2015/apr/30/how-to-end-hunger-lessons-from-the-father-of-indias-green-revolution?CMP=share_btn_tw.

40. Anthony Spaeth, "M. S. Swaminathan," *Time*, August 23, 1999, http://content.time.com/time/world/article/0,8599,2054394,00.html.

41. Swaminathan, "How to End Hunger."

42. Anil Agarwal, "From Begging Bowl to Bread Basket," *Nature* 281 (August 1979): 250–51, doi:10.1038/281250a0.

43. Norman Borlaug, "The Green Revolution, Peace, and Humanity," Nobel lecture, Nobel prize.org, December 11, 1970, http://www.nobelprize.org/nobel_prizes/peace/laureates/1970/borlaug-lecture.html.

44. "Percent of Employment in Agriculture in the United States (DISCONTINUED)," Economic Research, Federal Reserve Bank of St. Louis, last modified June 10, 2013, https://research.stlouisfed.org/fred2/series/USAPEMANA.

45. Dana Gunders, "Wasted: How America Is Losing Up to 40 Percent of Its Food from Farm to Fork to Landfill," Natural Resources Defense Council, August 21, 2012, http://www.nrdc.org/food/wasted-food.asp; and "The Food Recovery Hierarchy," Wastes, U.S. Environmental Protection Agency, last modified November 20, 2014, http://www.epa.gov/foodrecovery/.

46. "Wheat: Trade," USDA Economic Research Service, last modified January 24, 2013, http://www.ers.usda.gov/topics/crops/wheat/trade.aspx.

47. "Wheat," Agricultural Marketing Resource Center, last modified April 2012, http://www.agmrc.org/commodities__products/grains__oilseeds/wheat/.

48. Borlaug, "The Green Revolution, Peace, and Humanity."

49. Daniel Zwerdling, "India's Farming 'Revolution' Heading for Collapse," pt. 1, *All Things Considered*, NPR, April 13, 2009, http://www.npr.org/templates/story/story.php?storyId=102893816.

50. Chris Arsenault, "Small Farmers Hold the Key to Seed Diversity," *Reuters*, February 16, 2015, http://www.reuters.com/article/2015/02/16/us-food-aid-climate-idUSKBN0LK1PO20150216.

51. Ronit Ridberg, "Leading the Fight for Food Sovereignty, An Interview with La Via Campesina's Dena Hoff," *Nourishing the Planet* (blog), Worldwatch Institute, August 9, 2010, http://www.worldwatch.org/node/6514.

52. Daniel Zwerdling, "'Green Revolution' Trapping India's Farmers in Debt," pt. 2, *Morning Edition*, NPR, April 14, 2009, http://www.npr.org/templates/story/story.php?storyId=102944731.

53. Food and Agriculture Organization of the United Nations (FAO), "A Statement by FAO Director-General José Graziano da Silva" ("Lectio 'Aula Magna,'" lecture at the University of Gastronomic Sciences, Piedmont, Italy, March 25, 2013), http://www.fao.org/fileadmin/user_upload/newsroom/docs/2012-03-25-university-of-gastronomic-sciences-lectio-aulamagna-DG-speech-en.pdf.

54. Sean Gallagher, "The Poisoning of Punjab," Pulitzer Center on Crisis Reporting, September 2, 2014, http://pulitzercenter.org/reporting/asia-india-punjab-pesticide-food-production-pollution-chemical-agriculture; E. Blaurock-Busch et al., "Metal Exposure in the Physically and Mentally Challenged Children of Punjab, India," *Maedica* 5, no. 2 (April 2010): 102–10, http://www.ncbi.nlm.nih.gov/pmc/articles/PMC3150007/pdf/maed-05-102.pdf; and Neville Lazarus, "Pesticide Blamed for High Cancer Rate in India," *Sky News*, August 3, 2014, http://news.sky.com/story/1312160/pesticide-blamed-for-high-cancer-rate-in-india.

55. Harkirat Singh, "SGPC Goes in for Organic Farming on Land Attached with Gurdwaras," *Hindustan Times*, August 3, 2015, http://www.hindustantimes.com/amritsar/sgpc-goes-in-for-organic-farming-on-land-attached-with-gurdwaras/article1-1375798.aspx.

56. Lazarus, "Pesticide Blamed."

57. P. Sainath, "Farmers' Suicide Rates Soar Above the Rest," *The Hindu*, May 18, 2013, http://www.thehindu.com/opinion/columns/sainath/farmers-suicide-rates-soar-above-the-rest/article4725101.ece.

58. P. B. Behere and M. C. Bhise, "Farmers' Suicide: Across Culture," *Indian Journal of Psychiatry* 51, no. 4 (October–December 2009): 242–43, doi:10.4103/0019-5545.58286.

59. Max Kutner, "Death on the Farm," *Newsweek*, April 10, 2014, http://www.newsweek.com/death-farm-248127.

60. Keith O. Fuglie et al., *Research Investments and Market Structure in the Food Processing, Agricultural Input, and Biofuel Industries Worldwide*, ERR-130 (Washington, DC: USDA Economic Research Service, December 2011), 28, http://www.ers.usda.gov/media/199879/err130_1_.pdf.

61. Joseph Lofthouse, "Seed List," Garden.Lofthouse.com, last modified January 18, 2015, http://garden.lofthouse.com/seed-list.phtml.

62. Christy, "The Difference Between Open-Pollinated, Heirloom, and Hybrid Seeds."

63. "Carrots," Center for Urban Education about Sustainable Agriculture, accessed July 28, 2015, http://www.cuesa.org/food/carrots.

64. Renee Cho, "Improving Seeds to Meet Future Challenges," *State of the Planet* (blog), Earth Institute, Columbia University, February 8, 2013, http://blogs.ei.columbia.edu/2013/02/08/improving-seeds-to-meet-future-challenges/.

65. Michaeleen Doucleff, "Natural GMO? Sweet Potato Genetically Modified 8,000 Years Ago," *Goats and Soda* (blog), NPR, May 5, 2015, http://www.npr.org/sections/goatsandsoda/2015/05/05/404198552/natural-gmo-sweet-potato-genetically-modified-8-000-years-ago.

66. Tom Philpott, "Monsanto. Broccoli. I Love This. Really!" *Mother Jones*, July 17, 2013, http://www.motherjones.com/tom-philpott/2013/07/eastern-broccoli-monsanto-involved-plant-breeding-project-i.

67. Fuglie et al., *Research Investments and Market Structure*, 35.

68. Zwerdling, "India's Farming 'Revolution' Heading for Collapse."

69. Gary Paul Nabhan, *Coming Home to Eat: The Pleasures and Politics of Local Food* (New York: W. W. Norton, 2009), 27.

70. H. Charles J. Godfray, "Food for Thought," *Proceedings of the National Academy of Sciences of the United States of America (PNAS)* 108, no. 50 (December 2011): 19846.

71. Dan Barber, "What Farm-to-Table Got Wrong," *New York Times*, May 17, 2014, http://www.nytimes.com/2014/05/18/opinion/sunday/what-farm-to-table-got-wrong.html?_r=0.

72. GBSNP Varma, "Seed Savior," *Earth Island Journal* 29, no. 1 (Spring 2014), http://www.earthisland.org/journal/index.php/eij/article/seed_savior/.

73. Varma, "Seed Savior."

74. Varma, "Seed Savior."

75. Lynne Rossetto Kasper, "How Nikolay Vavilov, the Seed Collector Who Tried to End Famine, Died of Starvation," *The Splendid Table*, American Public Media, May 29, 2010, http://www.splendidtable.org/story/how-nikolay-vavilov-the-seed-collector-who-tried-to-end-famine-died-of-starvation.

76. Charles Siebert, "Food Ark," *National Geographic*, July 2011, http://ngm.national geographic.com/2011/07/food-ark/siebert-text/2.

77. Matt Bivens, "New Facts Point Up Horror of Nazi Siege of Leningrad: Warfare: The 900-day Blockade Was Lifted 50 Years Ago Today. Archival Materials Confirm Cannibalism." *Los Angeles Times*, January 27, 1994, http://articles.latimes.com/1994 -01-27/news/mn-15973_1_900-day-blockade.

78. Cary Fowler, "The Second Siege: Saving Seeds Revisited," *HuffPost Green*, August 18, 2010, http://www.huffingtonpost.com/cary-fowler/the-second-siege-saving-s_b _685867.html.

79. International Center for Agricultural Research in the Dry Areas (ICARDA), "World Heritage Seed Collection Makes Its Way from Syria to Svalbard," news release, April 22, 2014, http://www.icarda.org/features/press-release-world-heritage-seed-collection -makes-its-way-syria-svalbard#sthash.9wafkkj1.XlQD8Ytu.dpbs.

80. "Restoring Ancient Wheat," Israel Plant Gene Bank, last modified November 29, 2007, http://igb.agri.gov.il/main/index.pl?page=112.

81. "Restoring Ancient Wheat."

82. Christy, "The Difference Between Open-Pollinated, Heirloom, and Hybrid Seeds," blog, Seed Savers Exchange, May 2, 2012, http://blog.seedsavers.org/blog/open-pollinated-heirloom-and-hybrid-seeds.

83. "The Six Classes of Wheat," Wheat Foods Council, accessed April 20, 2015, http://www.wheatfoods.org/sites/default/files/atachments/6-classes-wheat.pdf.

84. "Wheat Classes," U.S. Wheat Associates, accessed April 20, 2015, http://www .uswheat.org/wheatClasses.

85. "History," Farm & Sparrow, accessed April 15, 2015, http://www.farmandsparrow .com/history/.

86. "About Me," Carolina Ground, accessed April 15, 2015, https://www.blogger.com /profile/13686531237848115048.

87. Eric Holt-Giménez and Raj Patel with Anne Shattuck, *Food Rebellions!: Crisis and the Hunger for Justice* (Cape Town: Pambazuka Press; Oakland, CA: Food First Books; Boston: Grassroots International, 2009), 6.

88. Frederick Kaufman, "The Food Bubble: How Wall Street Starved Millions and Got Away with It," *Harper's Magazine*, July 2010, http://harpers.org/archive/2010/07/the-food-bubble/.

89. Holt-Giménez and Patel with Shattuck, *Food Rebellions!*, 6.

90. Heather Wight, "The History and Processes of Milling," Resilience.org, Post Carbon Institute, January 25, 2011, http://www.resilience.org/stories/2011-01-25/history-and-processes-milling.

91. Wight, "The History and Processes of Milling."

92. Lauren Chattman, *Bread Making: A Home Course: Crafting the Perfect Loaf from Crust to Crumb* (North Adams, MA: Storey Publishing, 2011), 15.

93. "History," Farm & Sparrow.

94. Charles Spence et al., "A Touch of Gastronomy," *Flavour* 2 (February 2013): 4, doi:10.1186/2044-7248-2-14.

95. Spence et al., "A Touch of Gastronomy," 4.

96. John Prescott and Richard J. Stevenson, "Effects of Oral Chemical Irritation on Tastes and Flavors in Frequent and Infrequent Users of Chili," *Physiology & Behavior* 58, no. 6 (1995): 1117, doi:10.1016/0031-9384(95)02052-7.

97. Phipps Arabie and Howard R. Moskowitz, "The Effects of Viscosity upon Perceived Sweetness," *Perception & Psychophysics* 9, no. 5 (May 1971): 410, doi:10.3758/BF03210240.

98. Karen Hopkin, "Bread Texture Affects Salty Taste," *Scientific American*, December 25, 2013, http://www.scientificamerican.com/podcast/episode/bread-texture-affects-salty-taste-13-12-25/.

99. David Whyte, *Consolations: The Solace, Nourishment and Underlying Meaning of Everyday Words* (Langley, WA: Many Rivers Press, 2015), 221.

100. Tara Jensen, "Ramble On," *Smoke-Signals Baking* (blog), April 3, 2013, http://smokesignalsbaking.tumblr.com/post/47022494402/ramble-on.

101. Tara Jensen, "Baking Burial," *Smoke-Signals Baking* (blog), October 18, 2012, http://smokesignalsbaking.tumblr.com/post/33863980607/baking-burial.

102. Jensen, "Baking Burial."

103. Portions of this article and the Bread Tasting Protocol originally appeared in *The National Culinary Review®*, March 2015, published by the American Culinary Federation, and are reprinted here with their permission.

104. Derek Walcott, *Collected Poems, 1948–1984* (New York: Farrar, Straus and Giroux, 1986), 328.

OCTOPUS

1. Brian Wansink and Jeffery Sobal, "Mindless Eating: The 200 Daily Food Decisions We Overlook," *Environment and Behavior* 39, no. 1 (January 2007): 110–12, doi:10.1177/0013916506295573.

2. Caroline Scott-Thomas, "Most Eating Is Psychologically Motivated, Says IFST," *Food Navigator,* December 2, 2014, http://www.foodnavigator.com/Science/Most-eating-is-psychologically-motivated-says-IFST.

3. Palmiero Monteleone et al., "Hedonic Eating Is Associated with Increased Peripheral Levels of Ghrelin and the Endocannabinoid 2-Arachidonoyl-Glycerol in Healthy Humans: A Pilot Study," *Journal of Clinical Endocrinology & Metabolism* 97, no. 6 (June 2012): E917, doi:10.1210/jc.2011-3018.

4. Leif Hallberg, "Bioavailability of Dietary Iron in Man," *Annual Review of Nutrition* 1 (July 1981): 133, doi:10.1146/annurev.nu.01.070181.001011.

5. Henry David Thoreau, *Walden: An Annotated Edition,* ed. Walter Roy Harding (New York: Houghton Mifflin, 1995), 87.

6. Gidon Eshel et al., "Land, Irrigation Water, Greenhouse Gas, and Reactive Nitrogen Burdens of Meat, Eggs, and Dairy Production in the United States," *Proceedings of the National Academy of Sciences in the United States of America (PNAS)* 111, no. 33 (August 2014): 11996–2001, doi:10.1073/pnas.1402183111.

7. Jonathan Safran Foer, *Eating Animals* (New York: Little, Brown, 2009), 50.

8. Sylvia A. Earle, *Sea Change: A Message of the Oceans* (New York: Ballantine, 1996), 15.

9. Sylvia A. Earle, "Foreword," in *Spineless: Portraits of Marine Invertebrates, the Backbone of Life,* by Susan Middleton (New York: Henry N. Abrams, 2014), 12.

10. "Facts and Figures on Marine Biodiversity," United Nations Educational, Scientific and Cultural Organization (UNESCO), accessed April 27, 2015, http://www.unesco.org/new/en/natural-sciences/ioc-oceans/priority-areas/rio-20-ocean/blueprint-for-the-future-we-want/marine-biodiversity/facts-and-figures-on-marine-biodiversity/.

11. "Phytoplankton," Woods Hole Oceanographic Institution, accessed April 27, 2015, http://www.whoi.edu/main/topic/phytoplankton.

12. "Facts and Figures on Marine Biodiversity," UNESCO.

13. Boris Worm et al., "Impacts of Biodiversity Loss on Ocean Ecosystem Services," *Science* 314 (November 2006): 790, doi:10.1126/science.1132294.

14. Robert Krulwich, "The Hardest-Working Mom on the Planet," Krulwich Wonders, NPR, June 2, 2011, http://www.npr.org/sections/krulwich/2011/06/02/136860918/the-hardest-working-mom-on-the-planet.

15. Ed Yong, "Octopus Cares for Her Eggs for 53 Months, Then Dies," *Phenomena: Not Exactly Rocket Science* (blog), *National Geographic,* July 30, 2014, http://phenomena.national geographic.com/2014/07/30/octopus-cares-for-her-eggs-for-53-months-then-dies/.

16. Rachel Nuwer, "Ten Curious Facts About Octopuses," Smithsonian.com, October 31, 2013, http://www.smithsonianmag.com/science-nature/ten-curious-facts-about-octopuses-7625828/?no-ist.

17. David Whyte, "Heartbreak," David Whyte Readers' Circle, 2011–2012, DavidWhyte.com, http://www.davidwhyte.com/pdf%20files/Readers_Circle/Heartbreak.pdf.

18. David Whyte, *Consolations: The Solace, Nourishment and Underlying Meaning of Everyday Words* (Langley, WA: Many Rivers Press, 2015), 102.

19. Rebecca Solnit, "A Letter to My Dismal Allies on the US Left," *The Guardian,* October 15, 2012, http://www.theguardian.com/commentisfree/2012/oct/15/letter-dismal-allies-us-left.

APPENDIX I

1. "What Is Agricultural Biodiversity?," Convention on Biological Diversity, accessed August 12, 2014, http://www.cbd.int/agro/whatis.shtml.

APPENDIX II

1. Clem A. Tisdell, "Socioeconomic Causes of Loss of Animal Genetic Diversity: Analysis and Assessment," *Ecological Economics* 45, no. 3 (July 2003): 365–76, doi:10.1016/S0921-8009(03)00091-0.

FLAVOR GUIDES

WINE AROMA WHEEL

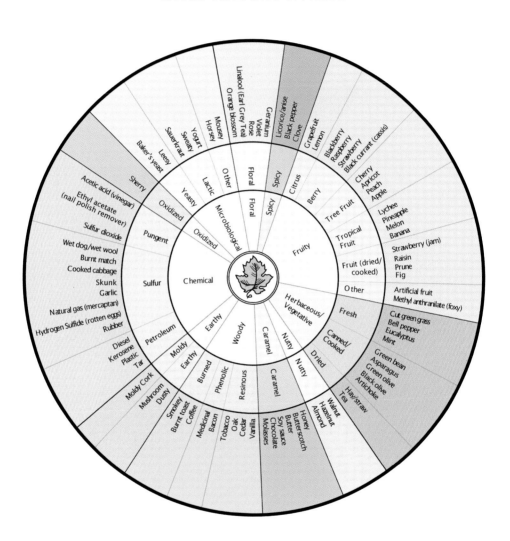

(The) Wine Aroma Wheel © 1990, 2002
A. C. Noble www.winearomawheel.com

CHOCOLATE FLAVOR WHEEL

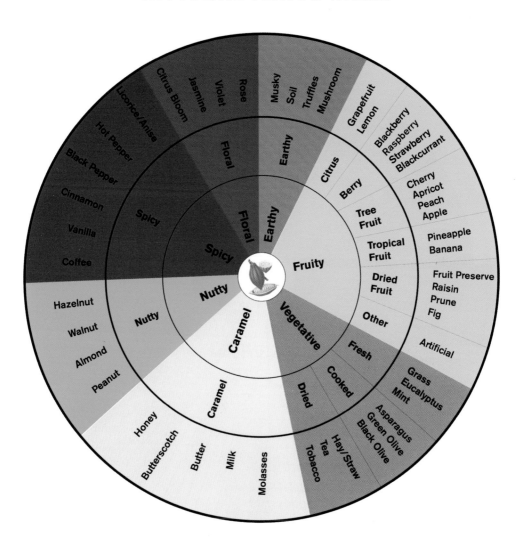

Bittersweet Chocolate Café via the C-spot® © 2006, 2011
Seneca Klassen www.c-spot.com/atlas/chocolate-flavor-profiles/

COFFEE FLAVOR WHEELS

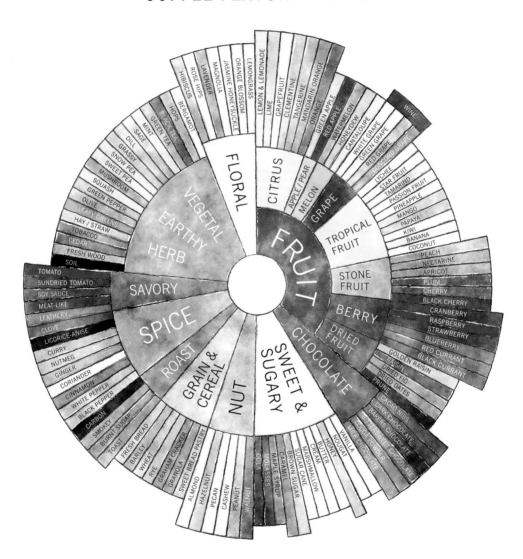

DEFECT WHEEL

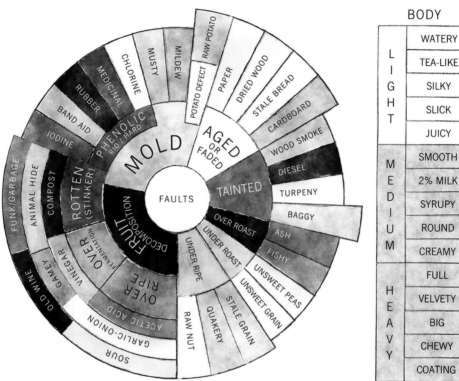

BODY

L I G H T	WATERY
	TEA-LIKE
	SILKY
	SLICK
	JUICY
M E D I U M	SMOOTH
	2% MILK
	SYRUPY
	ROUND
	CREAMY
H E A V Y	FULL
	VELVETY
	BIG
	CHEWY
	COATING

ADJECTIVES & INTENSIFIERS FOR COFFEE

MUTED DULL MILD	STRUCTURED BALANCED ROUNDED	SOFT FAINT DELICATE	DRY ASTRINGENT	QUICK CLEAN
CRISP BRIGHT VIBRANT TART	WILD UNBALANCED SHARP POINTED	DENSE DEEP COMPLEX	JUICY	LINGERING DIRTY

Counter Culture Coffee Taster's Flavor Wheel © 2013
www.counterculturecoffee.com/tasterswheel

COFFEE CUPPING FORM

A
SAMPLE

Dry coffee aroma (Circle any aromas that you can smell)
Flowery, fruity, herbal, chocolate, nutty, caramel, spicy, earthy, citrus, other:

Sweetness	**Acidity**	**Mouthfeel**	**Overall**
			Did you like it?

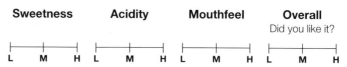

L   M   H   L   M   H   L   M   H   L   M   H

B
SAMPLE

Dry coffee aroma (Circle any aromas that you can smell)
Flowery, fruity, herbal, chocolate, nutty, caramel, spicy, earthy, citrus, other:

Sweetness	**Acidity**	**Mouthfeel**	**Overall**
			Did you like it?

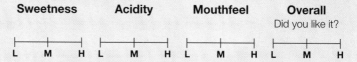

L   M   H   L   M   H   L   M   H   L   M   H

C
SAMPLE

Dry coffee aroma (Circle any aromas that you can smell)
Flowery, fruity, herbal, chocolate, nutty, caramel, spicy, earthy, citrus, other:

Sweetness	**Acidity**	**Mouthfeel**	**Overall**
			Did you like it?

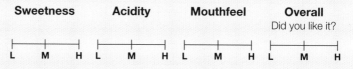

L   M   H   L   M   H   L   M   H   L   M   H

D
SAMPLE

Dry coffee aroma (Circle any aromas that you can smell)
Flowery, fruity, herbal, chocolate, nutty, caramel, spicy, earthy, citrus, other:

Sweetness	**Acidity**	**Mouthfeel**	**Overall**
			Did you like it?

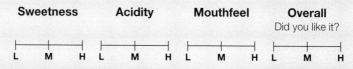

L   M   H   L   M   H   L   M   H   L   M   H

Seven Seeds Coffee © 2015

BEER FLAVOR WHEEL

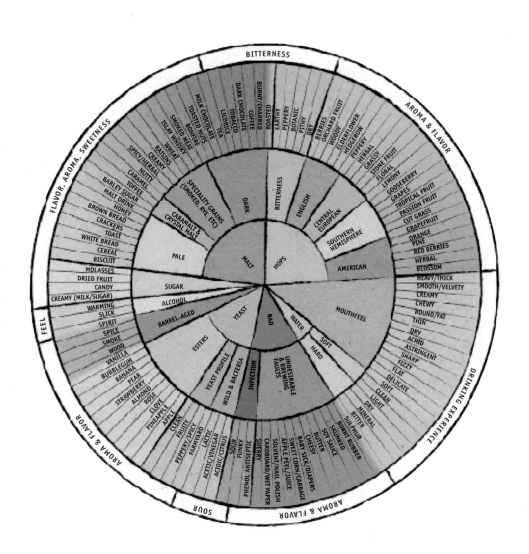

Mark Dredge, *Craft Beer World* © 2013

BREAD TASTING GUIDE

CRUMB

1. **SWEET**/ DAIRY
2. **SOUR** / FRUITY
3. **SOUR** / DAIRY

CRUST

4. **ROASTED**
5. **FRUITY**
6. **RESINOUS**
7. **TOASTY**
8. **SWEET**

GRAIN CHARACTER

SIMPLE   MODERATE   COMPLEX   OVER-FERMENTED

KEY

1. MILK, BUTTER, DIACETYL (POPCORN BUTTER)

2. GREEN APPLE, GRAPEFRUIT, LEMON, VINEGAR

3. FRESH CHEESE, BUTTERMILK, PLAIN YOGURT, AGED CHEESE

4. BAKED ONIONS, DARK BEER, BAKED CHESTNUT, CHEESE GRATIN

5. FIG, RAISIN, STEWED FRUIT

6. FRENCH ROAST COFFEE BEANS, VANILLA BEAN, AGED BALSAMIC VINEGAR

7. MALTY, POPPED GRAINS, NUTTY

8. BUTTERSCOTCH, TOFFEE, CHOCOLATE, MOLASSES

RAW STARCH, RAW GREEN BEANS, PEASHOOTS, STRAW, DRY YEAST

COOKED SPAGHETTI, STEAMED POTATOES, COOKED OATMEAL, YEASTY CHAMPAGNE

COOKED WHOLE GRAINS, COOKED DRIED BEANS, GREEN OLIVE, FLINT, SLATE, MINERAL

BEER, GRAPEFRUIT PITH, "TURNED" RED WINE, SHERRY

The Aroma & Flavor Notes for Bread Info-graphic (*"The Bread Flavor Wheel"*) and the Protocol for Sensory Evaluation of Bread (*"Bread Tasting Protocol"*) are the copyrighted property of Michael Kalanty and are used with the permission of the author. All rights are reserved and protected.

Michael Kalanty © 2015
www.michaelkalanty.com